HOW BIRDS FLY

By John K. Terres

SONGBIRDS IN YOUR GARDEN
THE AUDUBON BOOK OF TRUE NATURE STORIES
THE WONDERS I SEE
HOW TO ATTRACT BIRDS THE YEAR 'ROUND
DISCOVERY: Great Moments in the Lives of
    Outstanding Naturalists (editor)
LIVING WORLD BOOKS (editor)
THE AUDUBON NATURE ENCYCLOPEDIA (editor-in-chief)
HOW BIRDS FLY: Under the Water and Through the Air

HOW BIRDS FLY

Under the Water and through the Air

John K. Terres

HAWTHORN BOOKS, INC.
Publishers/NEW YORK

This book is dedicated to The Princess,
and to Everyone Who Has Ever Wanted to
Fly with the Birds in Their Adventures
Through the Air.

AUTHOR'S ACKNOWLEDGMENTS

I am grateful to Dr. Alexander Wetmore, former Secretary of the Smithsonian Institution, Washington, D.C., for critically reading the chapter on the evolution of birds and of bird flight, and for his editorial corrections and additions to the material. To Dr. Oliver L. Austin, Jr., I extend my thanks for reading the chapter on the flight of the albatross. I am especially indebted to Dr. Dean Amadon for critically reading the entire manuscript, for his many helpful and stimulating suggestions, and for the Foreword to the book. However, I take full responsibility for the work, and any errors that may appear in it are my own.

Migrant birds, cranes, for instance, fly
and fly, and whatever thoughts there may be
in their heads, great or small, they will
still fly . . .

Anton Tchekhov, *The Three Sisters*

CONTENTS

FOREWORD

"The way of an Eagle in the Air," was one of the mysteries that baffled Solomon, the Wise Man of the Scriptures. Today we know, at least in a general way, the adaptations that enable the bird to launch itself skyward: the flexible vanes, strong and tough, yet "light as a feather"; the great pectoral, or breast, muscles that operate the wings, sometimes for hours on end; and the hollow yet strong bones that further reduce the bird's weight. And we understand the properties of air and how, under proper conditions, it can support the bird, insect, or aircraft.

Nevertheless there is still much to learn. Aside from the engineering and technical aspects of flight, birds are beautiful, graceful creatures. When on the wing, they seem to symbolize man's yearning for freedom, independence, and escape from the more sordid aspects of existence.

John K. Terres has devoted a lifetime to patient observations of birds and other wildlife. In this book he enriches his own wide experience with a vast array of facts and episodes gleaned from the literature and from his years as the editor of *Audubon Magazine*. But above all he was inspired to study, and later to write about bird flight because he was fascinated by those masters of the air, the eagles and the falcons. A splendid peregrine falcon named The Princess, trained and flown by Mr. Terres, forms a focal point for his book *Flashing Wings*. It imparts to the account almost the suspense of a work of fiction. But the author ranges far afield to consider also the flight of hummingbirds, of albatrosses on the high seas, and, yes, of penguins through the water. The result is an account, charming, authoritative, and complete, of one of the world's genuine wonders.

DEAN AMADON, Chairman and Lamont Curator of Birds, Department of Ornithology, The American Museum of Natural History, New York.

AUTHOR'S NOTE

Because I watched her fly within a few feet of me almost every day for many years, The Princess, a trained falcon, appears now and then in this story. It is because I knew her better than any other bird of my lifetime that The Princess lives for me more than any other. Of the thousands of birds I have seen in some of their flying adventures told of in this book, each has taught me something about how, and why, birds fly. Through The Princess I first learned the meaning of the shape of a bird's wing to its flight. Of the approximately 650 species of native North American birds, many have wings adapted especially to their ways of life. There are groups of birds, like The Princess, that are swift-wings, others are slow gliders and soarers, others have wing shapes adapted to days, weeks, or even months of soaring about at sea, without any need to alight on land. Still others have rugged wings adapted to darting through forests or beating through underbrush, either in pursuit of their prey or to escape from their enemies.

Not all birds fly safely—some miscalculate and come to disaster; others have accidents in nature over which they have no control. A group of migrating birds, flying low over a New England mountain-top, are suddenly caught in a powerful downdraft. Unable to escape they are dashed against the mountain slope, where their broken bodies lie scattered as leaves before the wind. A prairie falcon stoops in a blazing dive at a jack rabbit, sweeps too close to the earth, and lies helpless with a broken wing. A golden eagle flies into a high-tension power line, and its great body is a crackling inferno, then a charred mass. A hurricane carries a seabird far inland. Forced by weakness and hunger to alight, it slowly starves, for it can fly up only from the surface of the water.

But for every accident there are millions of pairs of flashing wings that carry birds successfully through life, or at least until they have nested and have reared their young. For it is the driving purpose of birds, and of all the natural world, that each may produce others of its kind that they shall not disappear from the earth. Nowhere in nature do we see life gayer, more dynamic, or more meaningful of freedom than in the glorious, unrestrained flight of birds.

I had watched The Princess fly hundreds of times. It is to her that I owe the inspiration for my own interest in bird flight that has now spanned so many years. This is more than a part of The Princess' story. It is a story of some of the magic and mystery, the excitement and drama of birds that follow the skyroads of a vast ocean of air to all parts of the world.

JOHN K. TERRES

Chapel Hill, North Carolina

CHAPTER I

"TEACHING" A BIRD TO FLY

One sunny afternoon in May, I took my peregrine falcon, The Princess, from her perch in my backyard and carried her on my fist to the edge of the village for her daily flight. At the time I was living in a town in western Pennsylvania, and there was to be a special excitement about my adventure with The Princess that day. A local photographer had come along to take motion pictures of her in flight, and the whole village had turned out to watch.

I looked at The Princess perched quietly on my fist. Eagerly she had jumped to my gloved hand from her low block perch, all the while uttering a wild wailing like the squeaking of a rusty pump. She was hungry, and she was telling me she was ready and excited at the promise of a flight that would end with the food she craved.

Her black eyes glistened in her dark-feathered head. From the corners of her bluish, hooked beak, markings like the downsweeping ends of a black mustache led down each side of her face. Her bullet-like head was sunken between her broad shoulders, and her powerful back tapered downward to the long trim tail and pointed wings folded at her sides. Those long wings, belonging to one of the fastest birds that flies, reached to the end of her tail where the wing points crossed like a pair of opened scissors. The big yellow feet clutching my gloved fist were gnarled and strong, with each toe ending in a curved talon sharp as a needle.

This was my Princess, the noble peregrine falcon, as proud and fierce as the falcons that had been trained by kings and princes of the Middle Ages. Yet I had seen her in moments of gentleness when she crouched at my side like a pet dog, or flapped her wings eagerly at my approach to her perch. Our bond, which in the beginning had meant only food to the falcon, had become one of strong affection.

The Princess and I had been together for about a year, ever since a spring day when my friend Al Nye, an expert falconer who knew that I loved hawks, had brought her to me. She had been hatched and reared by the parent falcons, along with a brother and sister, high on a rock ledge of a cliff overlooking a river valley. The Princess was the last one left on the bare nesting ledge—her brother and sister had flown only a few hours before. She was several weeks old and fully feathered.

It had been up to me to train The Princess, but I really could not teach her anything about flight. She and her bird ancestors had practiced the principles of flying or gliding for about 140 million years, from the time the first reptilelike birds began to volplane from tree to tree. All that I could do was to give her the opportunity to develop speed and accuracy in flight through much practice.

Long before I had begun to train The Princess, I had heard of a dramatic experiment with domestic pigeons which told me that birds do not really need to learn to fly. J. Grohmann, a German scientist, raised some young pigeons in narrow tubes that prevented them from moving their wings. He allowed another group of pigeons of the same age to be raised by the parents in a nest in the normal way. These young pigeons, as they grew, could exercise their wings by vigorously flapping them while they were still in the nest, as many young birds often do.

When both groups of pigeons had developed to an age where their flying ability could be tested, Grohmann took the birds into the open and tossed them into the air. To the astonishment of many people who saw the test, the pigeons raised in the tubes flew away as strongly as those that had been unrestrained in the nest. Grohmann had proved that the instinctive behavior pattern of flight matures in a young bird at a steady rate regardless of its opportunity to practice flying.

Although birds have the ability to fly at the time they are ready to leave the nest, they still need lots of practice to develop skill in the air. My friend Al Hochbaum has for years raised and studied wild waterfowl at a refuge in Canada. He found that the instinct to fly matures in wild ducks at the time the primary or flight feathers

1.   *A peregrine falcon tethered to a block perch.*

of the wings of young ducks are fully grown. Although his young ducks flew instinctively, they had to learn to relate flying to their environment. They had to become skilled in alighting on the water and until they had practiced it often, they came down too heavily with a loud splash, or overshot the place where they wanted to alight.

The young ducks also had to learn how to use the wind when they were alighting or taking off. Whenever they landed downwind, or "with the wind under their tails," they crash-landed. It was only through experience that they learned to prevent this by facing into the wind when alighting.

In a way, I suppose I helped develop the flying skill of The Princess much as my high school track coach had trained me to run —by encouraging her to fly as often as possible. To keep her captive without causing her discomfort or harm to those stiff flight feathers of her powerful wings (they could drive her through the air at speeds up to 100 miles an hour or more), I put a short leather strap, called a jess, on each of her legs. To the jesses I attached an eighteen-inch-long leather leash and tied the other end to the base of her foot-high perch. This allowed The Princess room to exercise by jumping the length of her leash from her perch to the ground, and it also kept me from losing her.

All wild birds love to fly, but in the beginning I had to trick The Princess into her first practice sessions. Not that she was lazy —I simply had to supply an incentive to make her fly. I did it by taking advantage of her appetite. A falconer's training and control of his bird is through her hunger, and he never flies her free unless she is hungry. Hunger makes the falcon obey the falconer's wishes. She learns what he expects of her, knowing that she will be rewarded if she responds.

Even before The Princess had flown a foot, I had her jumping down from her block perch to the lawn to reach the baited lure I had flung on the ground before her. The lure was, in the beginning, a canvas-covered padded weight, heavy enough so that The Princess could not carry it away during her first flights. To the canvas cover of the lure, I had sewn thin leather strips to which I tied, for her

4

once-a-day meal, small pieces of fresh, raw beef. Every other day I alternated the lean beef with freshly killed poultry. Falcons, like hawks, are meat-eaters, and The Princess had learned, after feeding at the lure day after day, that the sight of it meant food to appease her hunger.

As The Princess became stronger, I coaxed her to make short flights from her perch to the baited lure, always keeping her on her leash and a long cord tied to the base of her perch. This would prevent her from flying away if a dog should rush into the yard during her flights or if some other object strange to her should appear before her suddenly.

Within a few weeks The Princess, still on her leash and long cord, was flying 200 feet or more from her perch to the lure. I knew then that the lure had such power over her that I could fly her free. Ever after that, it was the sight of the lure when she was hungry that drew her to me as though I had a magic spell over her. Flying about, high over the fields and woods, she would often turn in the air to watch me. If I wanted to recall her, the sight of the lure as I whirled it around my head would bring her streaking back.

Day after day The Princess grew stronger and more skillful during her practice flights at the edge of the village. Each day, more and more of the villagers came to watch and to gasp at the falcon's blinding speed as she coursed back and forth like a giant swallow low over the fields. Sometimes she rose into the wind and spiraled upward, climbing higher and higher with rapidly beating wings until she was a distant black speck in the immense vault of blue over-head. Then, when she seemed about to disappear, she would fall away in a long slanting dive toward a hilltop a quarter of a mile away. Just before crashing into the ground, she would nose up out of her dive and shoot back high into the air, like an arrow released from a bow. Cheers came from the crowd gathered at the roadside, then applause as The Princess came hurtling back to alight on the lure I had flung on the grass.

It was a splendid, a magnificent achievement, and The Princess seemed to glory in it. But beyond the physical fact lay something she could never tell me—the mystery of how she mastered the air,

5

of how she beat upon it to use it to her every need, as a powerful swimmer moves with grace and ease through water. I had always wanted to know The Princess' secret. Now on this May afternoon, perhaps the first motion pictures of her in flight would tell the story.

CHAPTER II

A FALCON CAPTURES HER PREY

As I walked across the greening pasture at the edge of the village that May afternoon, I had no idea that a double drama lay in store for us. The spring air was heavy with the fragrance of wild honeysuckle, and the distant woods were misted over with the pale green of new leaves. I had asked the villagers to remain at the roadside, and they had respectfully obeyed my wishes. I had explained that too many people near a falcon during her flight can confuse her and might keep her from returning at the moment I wanted to recall her. Only the photographer had moved out with me toward the center of the pasture, and he kept a discreet distance away, his camera poised for action.

I was holding The Princess on my gloved left fist, about level with my chest. With my right hand I released the swivel and leash from her short leather jesses, and she was free to fly unencumbered; however, the falconer never takes the jesses from the hawk's legs during her lifetime with him. They are the only means by which he can grasp the bird without touching her body, and can restrain her at the lure or at her prey on the ground. With a firm hold on her jesses he can recapture her and restore her to the leash.

A falcon would be frightened and resentful if her owner should pick her up bodily. This would be too much for her fierce pride, and she would likely fly away from him at his approach at the end of her next free flight, and he would lose her.

Slowly I raised The Princess high over my head and faced her into the strong westerly breeze. Birds prefer to take off into the wind and to face into it when alighting. While doing so they have greater

7

control over themselves in that critical moment when the wind might overturn them as they leave the ground or touch down for a landing.

The Princess leaned forward, her head low, her wings partly spread and stretched out from her sides. Suddenly she crouched, pressed down on my hand with her feet, and leaped into the air. With her first rapid wingbeat she was above my hand and rising into the wind. This is the way landbirds usually take off from a solid perch and are airborne. If a bird is on the ground, its leap carries it high enough that it can begin the first few horizontal wingbeats without striking its wings on the ground.

Using both her wings and feet, The Princess had sprung upward. Now she leveled off after her initial rise into the air. With rapid strokes, dipping her wings below her body in a deep downbeat, she sped straight away, about ten feet above the ground. As soon as she had left my hand, I reached into the bag slung over my shoulder. I grasped the lure hidden there but did not bring it out at once. I waited.

When The Princess reached the far end of the pasture and began climbing straight up into the sky, I shouted to her. Then I whipped out the lure and swung it in a wide circle around my head. The Princess saw it. She banked swiftly by dipping one wing far down. As she turned, I could see her tail tilted sidewise in the same direction as the downtilted wing. She had spread the tail feathers wide, like an opened fan, to help her make the turn.

Birds make remarkably good uses of the tail. Some use it to help turn over and over in flight (loop-the-loop), to fly upside down, and even to do backward somersaults in the air. Most birds will pump the tail up and down, as well as flap their wings, to help keep their balance when clinging to an unsteady perch. A pet male American sparrow hawk, or kestrel, I once owned would often perch on a slack overhead wire in my backyard, leaning forward and back, teetering precariously, but with his tail pumping up and down rapidly to help him hold his place.

The Princess had made her turn and was now approaching me at incredible speed, appearing to grow from small to large in seconds. As she came on, her dark wings like scimitars sliced the air at her sides. Her dark eyes looked straight into mine, and I saw her black mustache boldly marked against her light throat.

8

I swung the lure faster and faster. Suddenly she was upon me. I had timed the swing of the lure so that it was swinging back to me as she stooped with her feet outthrust to seize it. I jerked it away and into the grass. She swept past me and turned in a tight circle to come back. I picked up the lure and again she stooped at it as I swung it around my head. Again I twisted it away in time to elude her grasp. And so we played our training game together— The Princess flying like a black arrow back and forth over the length of the pasture, now with the direction of the wind, now against it, but always returning to me.

One day, while playing this game of wing-strengthening early in her training, The Princess suddenly tired of it. She was hungry and her patience was exhausted. I had thrown out the lure a number of times and she had stooped at it beautifully. To keep her on the wing a while longer, I had hidden the lure under my coat and was standing with my back to her as she approached, low over the ground. Instead of flying past me she swooped up in back of me and alighted on my head! The villagers watching from the roadside laughed in delight. When The Princess dropped down to my shoulder, I brought the lure into view and let her feed from it until she was full. That ended her exercising for that day.

The Princess had flown by me ten times and had made ten passes at the lure when I decided she had had enough of low flight. All the while I had heard the whirring of the photographer's motion picture camera, although I was barely conscious of it. I was too engrossed with The Princess. Now it was time to get her to rise up to a high pitch (altitude) from which she could make her most spectacular flight—the falcon's tremendous power dive from at least 1000 feet in the air.

I had planned it very carefully I thought. When The Princess reached a great height above the pasture, I would swing the lure. As she began her dive, I would throw the lure high in the air for her to strike as she reached the bottom of her long dive from the sky. With the photographer just behind me, we might capture it all on film and I would know exactly how The Princess managed it.

I hid the lure. The falcon, after circling me a few times, began rising higher and higher into the wind. As she rose, she alternated her quick wingbeats with a short sailing, in which her wings were held straight out from her sides. Higher and higher she went, her circles becoming smaller and smaller. When she was barely distinguishable in outline, looking no larger than a circling chimney swift, I knew that she had reached about 1000 or possibly 1200 feet above the earth. There The Princess "waited on," heading into the wind on rapidly beating wings, but remaining stationary above me. She was watching, expecting me to throw out the lure. It was the moment we both had waited for.

I heard a shout from behind me. I turned. Several of the villagers were pointing in the air. Three domestic pigeons, a white one in the lead, with two blue-gray ones slightly in back, were flying swiftly into the wind toward the far end of the pasture. Within seconds they would be below the point where The Princess hovered high in the sky. Apparently the pigeons did not see her as they continued their flight toward a barn on the distant horizon.

I forgot about the lure and watched The Princess. The pigeons were now directly under her. She turned over on her side and with a few powerful strokes of her wings began her dive. It was not until then that the pigeons realized that she was overhead. Perhaps they heard the wind rushing past her wings for they made a quick turn and, with the wind at their backs, headed for a woodland just beyond the pasture.

A pigeon is a swift-flying bird. I believe that with their own top air speed of 80 to 90 miles an hour, and with the 20-mile-an-hour wind carrying them, they were traveling at a ground speed of at least 100 miles an hour.

Fast as they were, they seemed to be standing still as The Princess closed the gap between herself and her prey. Like an air missile following a plane, she fell like a lightning bolt on the white pigeon. She had almost reached it when the pigeon in a sudden new burst of speed, flapped its wings and shot straight up into the air. Both falcon and pigeon were at the edge of the woods and seemed about to collide. As the pigeon bounded upward, The Princess, to avoid crashing into a tree at the edge of the woods, was forced to shorten her dive. She shot straight upward and, as she passed the pigeon, reached out with her feet. A puff of white feathers came away in her talons, then floated on the air. But the pigeon was safe. It had escaped, possibly by a fraction of an inch. A falcon will not pursue its prey into the shelter of a tree or bush, and the white pigeon had alighted on a branch of a tall tree. The two blue-gray pigeons had gone on and had disappeared in the woods.

Now The Princess seemed to have gone berserk. She climbed high over the woodland, circling wildly as though she hoped the pigeons would fly up again. Then she slanted away with the wind, turned into it again and began climbing higher and higher in the sky. I had brought out the lure and was swinging it around my head and whooping loudly. But The Princess ignored me. She was not ready to end her day. She wanted a successful flight against live prey.

She had moved a quarter of a mile away until she was a speck in the sky. There she hung over some distant field, waiting on. Had she sighted another flying bird somewhere beyond the horizon? If she stooped out of sight and made a kill too far away, I might never find her again. At that moment a new actor came on the scene.

From over a pinewoods ridge to the east a lone crow appeared. It was flapping slowly toward me, about 400 feet above the ground, and was quartering into the wind. As it came over the pasture, it lifted higher as it saw me and cawed hoarsely. The Princess was still waiting in the distant sky, with no indication that she had seen the crow.

The crow had reached a point over the center of the pasture. It was the farthest from any cover and now The Princess made her move. Again she turned over on her side and shot down in a long slanting dive. She had much farther to come this time, but she was pursuing a slower although craftier quarry.

As she gathered momentum, she pumped her wings in short strokes that seemed to double her speed. She dropped so fast that what had been an almost invisible speck became, in seconds, a hurtling streak of black. Her wings were half closed at her sides, her feet tucked back under her closed tail feathers. Her beak cut the air ahead of the dark eyes now blazing with fierce eagerness. Some years ago, an airplane pilot was diving his plane at 175 miles an hour when a peregrine shot rapidly past him in her stoop at a wild duck. The pilot estimated the falcon's speed at 200 miles an hour, and it was possible that The Princess was flying that fast.

The whistling whine of her wings as they cut through the air must have warned the crow. It suddenly folded its wings to its sides and dropped earthward. The crafty bird knew that its slow flapping flight could not beat the falcon to the woods. Its only chance for safety would be on the ground. There it could escape the falcon merely by getting under a bush, or by running along under a fence, or by ducking behind a pole. Falcons rarely if ever attack their prey on the ground unless they have disabled it. I had known of crows that had escaped from falcons in these ways, but this one hadn't a chance.

About 20 feet above the ground, The Princess struck the falling crow a tremendous blow with her feet and rebounded high in the air, prepared for another strike. But the crow fell limp and dead, a flurry of its black feathers falling after it. The Princess circled down and alighted on the dead bird.

I had to approach her carefully now. When the falconer "makes in" to his falcon after she has made a kill, he bends low, talks to her, and moves very slowly toward her. The Princess had begun to pluck the crow, tearing out black feathers some of which stuck out of the corners of her mouth, giving her a comical expression. As I neared her, I dropped to one knee, pulled a small piece of raw beef from my falconer's bag, and extended it toward her. She looked at the beef eagerly, and for a moment I thought she might leave the crow to come to me, but she did not.

As I approached her, bent low, I finally got close enough to reach out and almost put the piece of beef in her mouth. The Princess stopped plucking the crow's feathers, reached down, and took the beef from my fingers. As she bolted it, I grasped her jesses between the fingers of my left hand and lifted her, still on the crow, and stood up.*

Now I helped her pluck the crow, but still held her jesses. Slowly I reached back in my bag and brought out the lure. There were several pieces of beef tied on it. I put my hand over the

* Contrary to popular opinion, a falcon does *not* retrieve, or bring back the prey to the falconer.

12

crow and put the lure under The Princess' feet, at the same time dropping the crow to the ground. After one swift craning of her neck to see what had become of the crow (I was standing over it), she began tearing at the beef. While she ate I attached the swivel and leash to her jesses, and I had retrieved my falcon.

Of course I could have let The Princess eat the crow. But many falconers claim that the meat of a crow is sometimes distasteful to a falcon and that she might not pursue one again after tasting its flesh.

As I walked back toward the road with The Princess perched on my fist, I suddenly remembered my friend the photographer. The sun was setting and the villagers had gone home as soon as the flight ended. But the photographer stood there watching me as I approached. His camera was hanging from his right hand.

"How did you do?" I asked.

He swore, scowled, and kicked the ground ferociously.

"The camera jammed!"

"Oh well," I said. "We'll come out again and take some more pictures."

But we never did. The photographer moved away within a few weeks but not before I saw his films. They showed me swinging the lure, with brief but thrilling glimpses of The Princess as she flew toward the cameraman, then away like a black streak.

From the pictures I did not learn one thing about the art with which The Princess, or any other bird, flies. But I did discover what The Princess could do in the pursuit of wild quarry, and I am sure that she learned something from her two flights that day.

CHAPTER III

HOW A BIRD FLIES
(FLAPPING FLIGHT)

Not long after the attempt to photograph The Princess in flight, I went to a nearby city to a three-day meeting of ornithologists from many parts of the United States and Canada. I could stay for only one day, because I did not trust the care of The Princess to anyone but myself.

The work of one of the scientists that I met there for the first time was of great interest to me. He was not only an ornithologist who had spent many years observing wild birds, but he was at that time specializing in studies of bird flight through slow-motion photography. He also knew a lot about aerodynamics (literally "air motion") and the science of flight, about which I knew nothing.

After we had talked awhile, I told him about The Princess and our recent attempt to photograph her in flight. He was interested immediately. He had never seen a peregrine falcon close-up. Although he had seen a few wild falcons on his bird-watching expeditions, it was always from a distance. The science meetings were ending that day, and I invited him to see The Princess.

"I am expected to fly back home tonight," he said, "and I don't have my camera with me." He hesitated. "All right," he said. "I'll stay over another day and come to see you in the morning."

When he arrived at nine o'clock the next day, I took him immediately to the abandoned poultry house where I kept The Princess locked up at night. No dogs or night-roving cats, even if audacious enough to attack The Princess, could bother her inside the building. There she was also safe from a night attack by any of the large horned owls that occasionally came within the tree-lined village. The Princess was a match for any enemy while she was in the air, but held within a small area around her perch by the leash, she would have been almost helpless against the assault of any larger animal.

As I opened the door and my friend saw The Princess, his eyes lighted. He whistled softly.

"She's magnificent!" he said. "The most streamlined bird I've ever seen!"

The Princess sat quietly facing us on her perch. She was standing on one big yellow foot, with the other drawn up in her belly feathers. She sat without moving her body, but turned her black-mustached face curiously to follow the stranger as he walked around her.

"How big is she?" he asked.

"Twenty inches from top of head to end of tail," I said. "She weighs two and a quarter pounds; and her wingspread is about four feet. Of course," I added, "the female peregrine, or falcon, is larger than the male or tiercel, as the falconers call him."

He nodded. Now he noted the breadth and depth of her chest muscles and the thickness of the bend of The Princess' strong wings.

"That's where she gets the power for her great air speed," he said.

I did not reply but waited. I had started making quick notes on a large yellow pad. The man was an expert on bird flight, and I wanted him to continue.

"The principle of all flight," he said, "has been experienced by every kid that ever put his hand out of the window of a speeding car, boat, train, or other moving vehicle. The hand, a thin flat surface, tilted slightly upward, will be pushed up in the air through which it is speeding. In other words, air will support weight."

He nodded at the falcon. "The Princess, or any other bird, is a living airplane. She and her kind kept their secret a long time, and it wasn't until man developed the modern propeller-driven plane in this country that he learned how birds fly. When men first tried to fly, they built flying machines ('ornithopters') with mechanical wings that simply flapped up and down, and were powered by

16

human muscle. Of course they didn't get off the ground because that is not the way birds fly—also, the human body is too heavy and unstreamlined to fly."

He paused and lit a cigarette. The Princess turned her head far over to one side to watch the blue smoke curl up to the ceiling.

"Beginning in the Middle Ages," he continued, "a long succession of 'tower jumpers' trusted their lives to home-made wings. What they didn't understand was the amount of air lift required to support a human body."

He turned again to The Princess. "Will she allow you to touch her breast muscles?"

I grinned. "I do that every day to see if she is in condition to fly."

"How deep would you say her breast is—that is, how far does her breastbone project below the axis of her body?"

As I knelt beside her and reached slowly toward her chest, The Princess spread her wings and opened her mouth, but she did not jump from her perch. Carefully I felt of her powerful breast muscles (the pectorals) that were layered on each side of her sharply keeled breastbone, or sternum.

"About five inches across and three to four inches deep," I said. The Princess reached down with her hooked beak and pinched the back of my hand. It did not hurt, but it showed her disapproval.

"All right!" he said triumphantly. "Here we have one answer to why a man can't fly. The human body is not streamlined as a bird's is, and it would need an incredible amount of muscle to drive it through the air. A person weighing a hundred and fifty pounds would need to have a breastbone projecting six feet outward. Six feet of sharply keeled breastbone like a bird's in order to accommodate muscles large enough to flap wings capable of lifting him from the ground! And it wasn't until 1680, after many centuries of tower-jumping casualties, that anybody pointed this out."

He paused thoughtfully. The Princess puffed out her feathers and shook them vigorously back into place. Then she lifted one big foot and with the long middle toe scratched one side of her face. Years later I saw a wood stork do that while it was flying over a Florida swamp. I had often seen The Princess "rouse," or shake out her feathers while flying, and John James Audubon in his account of the man-o'-war bird, or frigate bird in Florida, wrote that it often scratched its head while on the wing and because it dropped low in the air while doing so, proved an easy target.

My friend spoke again. "We've talked about The Princess' 'motor.' Now let's have a look at her wings. It is in their design and her use of them that we have the answer to a puzzle that took man a long, long time to unravel. Had he understood from the beginning that a bird does not fly merely by flapping its wings up and down, he might have discovered the secret of flight long before he did."

At that moment the morning sunshine pouring through a window touched The Princess' back. She spread her wings to the sun, holding them out almost straight from her sides.

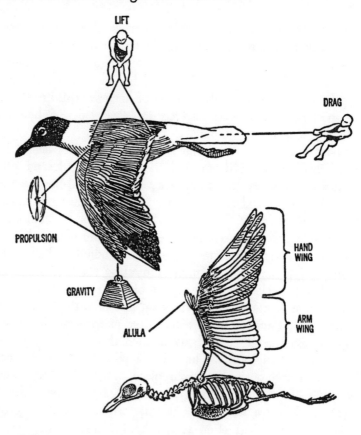

2. *The flapping flight of a laughing gull.*

"Those beautiful wings," he said, "are not only handsome to see but have a beautiful two-part function. The ten stiff outer primary feathers—from the 'wrist' out, or past the bend of each wing—are the 'hand' part of her wings. The slightly pointed primaries of the hand wing are the 'propellers' with which she flies, and they are operated from her wrists, not from her shoulders.

"When she flew for you the other day, as she bent the 'hand wings' in those deep strokes, she was pressing the air with each primary feather, twisting and adjusting them to drive her ahead as a propeller drives an airplane through the air. At the same time those almost straight feathers of her inner half of the wings—the secondaries—were horizontal or slightly tilted upward. They were acting as the wing of a plane acts to hold it up in the air, and they are moved or operated by a bird's shoulder joint. Every bird has a pair of propellers," he concluded, "and they are best seen in action through slow-motion photography of a bird in flight."*

He looked at his watch. "I've got to start back soon," he said. "Any questions?"

"At least a million," I said with a smile.

"One more thing," he said, "and I will be on my way. The ocean of air in which birds fly is a very real ocean. Like any other fluid it presses against every surface of anything submerged in it, with a pressure of 14.7 pounds per square inch of body surface at sea level, and less and less the higher we go.

"The air has become one vast highway for man riding in his aircraft, but we must remember that it has always been a skyroad for The Princess, for a pigeon, a crow, or for a wild goose or any other flying bird. It is the travelway over which birds can move to any place on earth; a haven from their enemies of the ground; a source of all their companions. It is a realm of all the storms that affect birds; a place where they can gather food by day; a world of moonlight and stars that light their night migrations; a dark hunting grounds for whippoorwills and owls. And it is the home of the wind that can carry birds backward in a hurricane, or can help speed them more swiftly to their summer or winter homes."

He stopped and smiled. "I seem to have gotten carried away with

* For a detailed description of bird flight, see in the Appendix, under Some Technicalities of Bird Flight, the section, *How a Bird Flies.*

my enthusiasm for bird flight. You'll understand when you begin to study it more seriously. Watch for the different kinds of flight. Study birds as they fly and get to know them by their flight patterns as you would know a friend from his walk."

We shook hands and he left. I never saw him again, but I did not need to ask him the "million" questions I wanted to ask that day. He wrote a book about bird flight, and it helped me ever after that to understand what I was seeing in my flight adventures with birds.

CHAPTER IV

ADVENTURES WITH VULTURES
(SOARING AND GLIDING FLIGHT)

When I was a boy, a pine forest grew on a hill overlooking the village where I lived in southern New Jersey. The villagers called the forest "Horse Heaven," because it was a dumping grounds for dead horses that were hauled there in a cart with high-built sides. That was long before automobiles were common, and most vehicles were horse-drawn.

We lived in farming country where there were many horses, mules, cows, and pigs. Some of these animals would get sick and die, and then the high-sided cart would haul them through the village streets. As the cart drove by, we youngsters could see the feet and hoofs of the larger dead animals sticking up stiffly above the sides of the cart. We called the cart "the dead wagon" (a wagon for carrying the dead). I remember that I shuddered whenever I saw it pass.

There were no sanitary laws in those days, in our community, compelling the burial of diseased animals. To prevent the spread of the dreaded anthrax and cholera, the dead wagon hauled them a safe distance from the village. There the turkey vultures fed on the carcasses and quickly disposed of them. Sometimes, from a distance, we boys watched the cart winding up the sandy road that led into the pine forest. Always in the skies over "Horse Heaven" hung large black turkey vultures on uptilted wings that they held in a flattened V as they glided or soared about.

Later, in my study of bird flight, I was to learn that this up-tilted position of the wings of soaring birds is called "dihedral." It keeps a forward-gliding vulture so well balanced in the air that it

3. *Turkey vultures. Note the flattened "V" of the wings as they soar and glide.*

does not roll over like a floundering ship in rough seas. Its body and wings simply teeter a little from side to side, resembling a man balancing himself on a tightrope with his arms outstretched.

The body of a bird is compact with the heavier parts grouped closely around the center of gravity and set below the horizontal line of its wings. The effect of the weight of the bird during its flight, as it is "suspended" between its wings, is somewhat like that of a pendulum. When gliding, vultures and other soaring birds often hold their wings at the dihedral angle to increase their stability.

The turkey vultures, or "buzzards" as they were called locally, were the most graceful soaring birds of any I had ever seen. I learned to recognize them instantly by their V-shaped flight. While they were still a long way off, I could distinguish them from a soaring hawk which holds its wings flattened.

The turkey vulture, with its six-foot wingspread, along with the golden eagle of seven-foot spread, the California condor of a nine- to ten-foot wingspread, and the red-tailed hawk with a four-and-a-half-foot spread, are some of the typically highly successful soaring birds. They are nature's most superb gliders over land and they live especially in open areas. There they soar about high in the sky and keep aloft mainly by floating on rising air currents.

These birds have evolved a type of wing that one scientist, D. B. O. Savile, calls the Slotted-Soaring, or High-Lift Wing. With their wings outspread, they soar in spirals higher and higher into the sky on warm updrafts of air called thermals. These are caused by the uneven heating of the earth's surface. The air over cities or bare fields heats more quickly than the air over forests or bodies of water. Because the warm air expands, and is lighter than cool air, it rises above the cooler air in slender columns or enormous bubbles. The soaring vultures and other birds seek out these thermals on which to gain altitude from which they can look down over the countryside in their search for food or for whatever joy or exhilaration that soaring may give them.

There was a strange fascination about "Horse Heaven" that both drew and repelled me. I had been interested in birds from the time I could first walk, and I had never had a close look at a vulture on the ground. At last my curiosity about vultures overcame my dread of the place. One day, with another boy, I walked up the sandy road and into the dense pine forest. I still recall vividly what we saw when we rounded a bend in the road. About fifteen or twenty big turkey vultures with flapping wings that seemed to us to spread at least ten feet across, were moving about over the carcass of a dead horse. Their heads and necks, like that of a turkey, were covered with dull red skin instead of feathers. Their heads were lowered, and with their strong, blunt, but hooked beaks they were devouring the animal.

When the vultures saw us they scrambled about in confusion and leaped awkwardly from the carcass. Almost stumbling over each other in their haste to get away, they hurried a few clumsy steps.

Then with a silky rustling of their feathers and a tremendous labored flapping of their wings, they got off the ground.

As they flew away and gained height, they turned and came back to glide by and look curiously down at us. We ran into the woods, fearful that they might attack us. Had we known vultures better at that time, we would not have been afraid. They were probably only anxious that we should leave so that they could get back to their feeding. About two years later, when I raised a young vulture, I smiled to think of my fear of those birds on that day.

The feet of turkey vultures are clumsy and weak, the claws dull and not able to seize and kill live prey as do the eagles, hawks, and owls. Adapted only to eating animals that are dead, turkey vultures are members of nature's wild sanitation corps. Scientists have named the American vulture family the Cathartidae, from a Greek word meaning "cleanser" or "scavenger."

Since those turkey vultures of my boyhood I have seen thousands of them in both summer and winter skies. Some were moving into the wind, others across the wind, and others with the wind. Almost always they moved without a wing flap, soaring in that effortlessly lazy fashion that distinguishes a vulture from most other birds. And always there was one particularly puzzling thing about their flight. I noted it especially one September day over a farm in North Carolina.

I was walking rapidly along a road through open fields, facing the wind. I looked up as I often do to search the skies for birds and saw ten turkey vultures behind me. To see one or two, or even three or four turkey vultures gliding along together is not unusual, but to see ten is extraordinary. They were spread out and facing into the wind just as I was. They were about 200 feet above the ground and moving steadily ahead. The vultures were in a low-altitude hunting flight, and as they scanned the earth below them, they paid no attention to me. Turkey vultures hunt their food by their wonderful sense of smell and keen eyesight. Dead animals are not always plentiful, and so they must keep flying to find new sources of food. When they overtook and passed me, I saw how they managed to move ahead faster than I, and without flapping their wings to do it.

Flapping means using energy for all birds that fly. To vultures that depend on sustained flight to keep them on an almost constant aerial search for food, it is important to conserve every bit of energy possible, especially when food is scarce. They do it by gliding.

The ten vultures were gliding against the wind. They were inside an air mass that was moving *against* their direction of flight (the wind would bring them odors of dead animals on the ground ahead of them), yet they moved steadily forward. They did it by occasionally diving toward the ground. The weight of their bodies, or the pull of gravity, gave them forward speed on the shallow dive. At the bottom of each long downward glide, they tilted the front edge of the wings upward (raised the "angle of attack," in aeronautical language) which pulled them out of the dive and shot them up and forward into the sky to about the height from which they had started. It was as though the vultures were riding a toboggan, a skillful way of moving ahead as a boy would ride his sled down a hill and with his momentum, glide up the next slope. It was a dramatic example of dynamic soaring, and of using gravity to move forward against the wind's direction.

I saw this clearly as the turkey vultures passed. But what puzzled me most about their flight was that, while heading into the wind, and sometimes across it, and briefly, with the wind, not one of their feathers was ruffled or out of place. Later, an aeronautical engineer to whom I posed this puzzle, explained it.

"A bird," he said, "whether it is soaring, gliding, or flapping, is suspended in the air, just as an airplane is in its flight. It is immersed in the air, just as a swimmer is immersed in liquid when he is in the water. If you were to drop into the still waters of a lake and began to swim across it, you would be like a bird flying in still air. You would feel the pressure of the waters all around you, but you would feel it a little stronger against your chest as you propelled yourself through it.

"Now," he continued. "If the waters of the lake in which you are swimming were suddenly transformed into a flowing river or creek, as you know, you could still swim in any direction. The pressure of the water would still be all around you. But as you swim, no matter whether against the tide or with it, the pressure against your chest would still be strongest. Although you are moving

25

through the water, you are suspended in it, just as a bird is suspended in the air, with almost equal pressure all around your body no matter in which direction you swim."

"That seems clear enough," I said.

"Let me give you another example," he said.

"Suppose you were swimming upstream against the tide of the downflowing current of the creek or river. You would have to swim against the current faster than its speed to make any noticeable headway to a person standing on the bank watching you. However, if you swam just fast enough to equal the flow of the current, you would appear to be standing still to the person on the shore. This is comparable to the way a gull, or a kingfisher, or a sparrow hawk flaps its wings and hovers against the wind to hold its place above one spot over the water or over the ground when it is hunting.

"Now," he said, "suppose you swam across the current toward the shore. You would probably need to angle slightly upstream to counter the current and to bring you to a point on the bank where the person is watching you. That is what a bird would need to do when it slants into the wind to reach a spot on the ground some distance away for which it is aiming. We say of the bird that it 'quarters,' or 'angles' into the wind.

"If you turned and swam with the current, you would travel faster than you did upstream or across stream. This is because you have the speed of the water of which you are a part to help speed you along in addition to your own swimming efforts. This is comparable with what happens to a bird when it decides to fly with the wind, or as we say, 'with the wind under its tail.' Its own gliding momentum or flapping speed is aided greatly by a strong wind, but it is still suspended in the air on which it is riding.

"That is why," concluded the aeronautical engineer, with a smile, "the wind does not ruffle a bird's feathers while it is in flight."

Not long after the engineer talked to me, I had an experience that showed me that if the wind does not ruffle a vulture's feathers while it is in flight, the attacks of another bird can.

One day The Princess, during one of her training flights over an open field, rose to about 1000 feet above my head. There she "waited on," with her wings beating rapidly but holding her position directly over me. I was "lure-hawking" that day. The Princess was waiting for me to swing the canvas-covered baited lure about my head to which she would come down in tremendous power dives that helped strengthen her wings and develop her aerial skill.

At that moment a large black turkey vulture soared across the pasture about 100 feet in the air. It was scanning the ground below it, obviously looking for a dead animal on which to feed. When it got directly below The Princess I was astonished to see her turn on her side, pump her long pointed wings a few times and shoot down on the vulture like a bolt of black lightning. At the bottom of her stoop, as she rebounded upward, she whacked the poor old vulture so hard that it rocked crazily in the air. A puff of black feathers exploded from the vulture, and it began to flap desperately in the direction of a distant woods. The Princess continued her attacks and struck the vulture twice more before it finally disappeared in the safety of the trees.

I was puzzled by The Princess' assault on the vulture. This big bird is not the prey of a wild peregrine. Perhaps she had struck the vulture in a spirit of play. I had once seen a wild peregrine dart at a kingfisher repeatedly, making it dive under water until the kingfisher almost drowned, without ever seriously trying to kill the bird. Possibly, The Princess may have resented the vulture's intrusion in the air over the field which she probably considered *her* flying territory. Birds are strongly possessive of their home bases, and even the smaller birds—kingbirds, blackbirds, and tiny hummingbirds— will rise into the air to attack birds many times their size if their nesting territories are invaded.

It was because the turkey vulture is one of the most graceful soaring and gliding birds in the world that I have admired it most of my life. The more southern black vulture, which I got to know better when I went to live in the South, does not confine itself to feeding on dead animals as the turkey vulture does, but occasionally attacks and kills young pigs, skunks, calves, and other live animals. For years I had been puzzled by the fact that the aggressive black vulture, which I had seen attack turkey vultures and drive them from dead animals along southern highways, was limited to the South. One day I learned the answer.

In 1949, August Raspet, a scientist who was studying the biophysics of bird flight, designed with a companion a motorless sailplane in which he hoped to soar in the air with vultures to learn some of the secrets of their flight. In order to follow birds in their short rapid turns in the air, he designed the glider for low sinking speed, or great buoyancy, and for slow speed and easy maneuverability. After the plane was built, Raspet equipped it with a small radio transmitter to send his scientifically recorded data to a recorder on the ground.

After being towed to a great height by an airplane and released, Raspet was guided to soaring vultures by radio reports from observers on the ground. Then he would descend to the altitude of one of them and follow it closely, observing it from 15 to 30 feet away.

From his flights with vultures, Raspet discovered that the turkey vulture has a slightly greater aerial buoyancy—less sinking speed—than the heavier and shorter-winged black vulture. But it was this slight difference, he believed, that accounted for the tremendous difference in the distribution of the two birds. The more buoyant turkey vulture lives over most of the United States north to the cool, forested regions of southern Canada; the aerodynamically heavier black vulture is limited to the southern part of the United States and the American tropics. Perhaps the hot sunshine there generates more of the vigorous rising currents of warm air (thermals), on which the heavier black vulture depends, than in the North.

Vultures usually do not shy away from the intrusion of gliders on their thermals as eagles and some of the larger hawks do. Phillip Wills, an English glider pilot, often went soaring in company with vultures. Wills had noticed that vultures, while gliding about, were always watching each other for a sign of descent that would suggest that one of them had discovered food. He also noted that when a vulture found a thermal, circled in it, and began to rise, neighboring vultures immediately followed it until a pyramid of birds was climbing in the air.

One day, Wills glided to one of the thermals occupied by a rising column of vultures and joined them. To his surprise, they accepted him as one of their soaring companions. Later, when he discovered a thermal and began to rise in it, the vultures followed his sailplane and climbed after him high into the air.

It was a series of events beginning more than thirty years ago that suggested to me that vultures may be surprisingly quick to learn new tricks in flight.

In 1935, Edward A. McIlhenny, a well-known bander of wild ducks in his wildfowl refuge in Louisiana, saw a black vulture in a remarkable maneuver that he and other ornithologists were unable to understand.

One day, while banding ducks, McIlhenny heard an airplane approaching. Glancing up he saw what he thought were two airplanes, one a little in advance of the other. As they came over his head, he was astonished to see that the second object was a living bird—a black vulture. It was gliding slightly above the airplane and about 200 feet in back of it. The plane was a U. S. Postal Department craft that carried mail between Houston and New Orleans, and it had a cruising speed of about 127 to 160 miles an hour. The remarkable thing was that the vulture was able to keep pace with the presumably faster flying plane; the puzzling thing was why it was following it so closely.

Two years later, McIlhenny saw, at the same place, the same drama re-enacted—a black vulture gliding close in the wake of the mail plane. A year later, McIlhenny's report of his unusual observations was published in a scientific journal.

But twenty years were to pass, long after McIlhenny's death, before a scientist was able to interpret what McIlhenny had seen. It was a trick of flight known to wild geese and swans for possibly millions of years, that the black vulture may have learned from following the plane.

When flying, each bird creates behind it a small area of disturbed air, the so-called "slip stream." Any bird flying directly behind another would be caught in this turbulent air. It would probably be canted over and thrown about in the eddies, thus disrupting its flight. Of the many kinds of birds that fly in flocks, most of them have probably experienced this. Some, however, have learned to use it to their advantage.

As each bird flies, some air is lost or spills over the wing tips causing a loss to the bird of "lift." This circulation of air creates an enlarging spiral wing-tip vortex behind each wing tip, with up-swelling air on the outer side of each wing.

4. *Canada geese in "V" formation.*

Wild geese and swans, by flying in formation, use this upswelling air to save their energy when flying. In the V formation of Canada geese for example, each bird flies, not directly behind the other, but aside or above the bird in front. By so doing each bird rests its inner wing tip in flight on the rising vortex of air from the bird's wing in front of it. By formation flight, a substantial power lost at each bird's wing tips is salvaged and used by others in the flock.

"This unquestionably," wrote D. B. O. Savile, the scientist, "explains McIlhenny's . . . observations of black vultures sailing behind aircraft."

The black vulture's ability to learn had given me one more example in my studies of bird flight, of the adaptability of birds—of adjusting their flight to their needs. It was another example of their capacity for making life easier for themselves.

CHAPTER V

THE GREAT GLIDER HIGHWAY (GLIDING FLIGHT)

Years ago, to me and to some of my bird-watching companions who also had not been there, Hawk Mountain, near Reading, Pennsylvania, seemed a mythical place. We had heard almost unbelievable stories about it. Someone said that watchers there, in one autumn day, could see hundreds of migrating hawks and many eagles, some of them passing so near that one could reach up and almost touch them. Even an occasional big black raven glided slowly by and once, a very rare white gyrfalcon had sped southward along the mountain ridge. I had never seen a raven or a gyrfalcon in the wild, and in those days, with my bird watching crowded into an occasional weekend, I felt lucky if I saw fifty hawks in a year.

It was said by ornithologists that the multitude of hawks, eagles, and vultures that passed southward along Hawk Mountain each autumn, had used that migration route for thousands of years. For a long time there had been a mystery about why the birds used it, but when scientists began to study the aerodynamics of bird flight, and to trace the migration route, the reason became clear.

The Kittatinny Ridge, of which Hawk Mountain is a part, begins 150 miles to the north in southeastern New York State. Southward at Hawk Mountain, the ridge suddenly narrows high above the rolling Pennsylvania farmland. It was there that hawks, eagles, and other gliding birds that followed the ridge southward, were brought closely together in their parallel flight. With sensitivity in their wings, tails, and bodies for taking advantage of every current of air favorable to their long migration flights, the birds followed the ridge for its

5. *The Great Glider Highway—from the Lookout rocks of Hawk Mountain, Penn.*

uplifting currents of air that created their invisible but supporting glider highway.

Unlike the occasional thermals, or vertical columns of warm air rising from fields heated by the summer sun, the autumn glider highway is formed from horizontal currents of air (winds) that are bounced off solid objects and deflected upward. Scientists call these obstruction currents. They are also created when steady or prevailing winds strike low hills, buildings, sand dunes, and even waves and the sides of ships at sea. Shearwaters, petrels, and other small or medium-sized oceanic birds glide on air currents bounced up from waves. Gulls often glide lazily close to a ship, scarcely moving their wings. They are riding the rising air currents created by winds striking the sides of the vessel as it moves across the water.

Only on certain days in fall at Hawk Mountain and in certain weather does the invisible glider highway exist. It is most attractive to birds on clear days when strong north and northwest winds strike the flanks of the ridge, and especially powerful currents of air are swept upward.

When the wind is not blowing, the hawks may not fly at all. Or, they may be spread out over the Pennsylvania countryside, using the thermal air currents of the open country to lift them so high as to be sometimes invisible to the watchers below. This is called by scientists *static soaring*, because the birds are lifted with little or no effort of their own by vertical masses of air. Carried to a tremendous height by a thermal, the hawks then turn and glide southward to continue their migration flight. Thermals, when they are available, are used by them to gain height for a long descending glide forward. As I watched them, I almost seemed to hear the birds say: "The journey is a long one. Why 'walk' (flap) when we can ride?"

It was a golden day in early October when I first saw Hawk Mountain. I had left The Princess safely at home on her perch, and with several college students who were also interested in hawks and eagles had driven the 200 miles from my home to the mountain near Hamburg, Pennsylvania. At the village of Drehersville, we had turned up the road that leads over Hawk Mountain. Near the top of the wooded ridge, at the beginning of a footpath by the side of the road, we parked the car. Then we climbed the trail which led through a dense oak woods.

Abruptly the path ended at the edge of the woods. Before us lay a huge pile of whitish boulders that were scattered about as though some giant had flung them there. Standing on the rocks, we looked almost straight down into the valley of the little Schuylkill River, 1000 feet below. On either side of the ridge, long valleys stretched away in a patchwork of fields and woods. We stood as though on the prow of a ship, with the ridge dropping sharply from in front of us to tree tops 100 feet below. From there, gradually sloping upward and away, our ridge extended northeastward to broaden into a mountain of five peaks at eye level about a quarter of a mile away.

It was over this wooded mountain ridge that we saw the first moving specks in the sky that were approaching hawks and eagles. They did not travel as a close-knit flock but were spread out like a loose squadron of warplanes. Each hawk and eagle, although in a group, was still an individual, occasionally circling and mounting to gain altitude, then gliding ahead on the air currents bounced up from the slopes below by the brisk north wind.

As the specks grew larger, our excitement grew. They were moving rapidly toward us, and we began to recognize the leading ones as hawks. The first, a small, sharp-shinned hawk, had short rounded wings with the "fingers" or flight feathers spread and a long, squared-off tail. It flapped three times, then sailed, flapped three times again, then sailed. Its flight was a rhythmic *flap, flap, flap,* then a slow *g-l-i-d-e,* timed just about as one might speak these words.

The sharp-shinned hawk is not adapted as much to soaring as the larger hawks we were to see this day. It does not have as large a wing surface or supporting surface in relation to its weight as have eagles, vultures, and larger hawks. It must flap frequently, even on the updrafts of air, to keep its altitude and to stay aloft. It is a bold, fierce little hunter, with a long tail and tough, rounded wings adapted to tremendous bursts of speed over a short distance. It uses its long tail as a rudder for quick turns as it recklessly dashes in and out of woods and thickets. When hunting, it depends on surprising its victims. As it appears suddenly in a forest clearing or over a hedge,

34

it may snatch a small bird out of the air with its sharp talons, or pluck one off the ground in full flight before it can escape.

More sharp-shins were in the air over our heads. Some of them swept up from the ridge below us. When they skimmed low over the rocks, for a split second we were face to face with them. Then they lifted sharply over our heads and were gone. A Cooper's hawk came toward us, flapping and gliding, with much the same kind of flight as a smaller sharp-shinned hawk that was passing by. The larger Cooper's hawk also has the short wings and long tail like its close relative the sharp-shinned, which gives it control of its dashing flight as it darts in and around trees and thickets. My friend Herbert L. Stoddard of Thomasville, Georgia, an expert on the bobwhite quail and other birds, discovered that the Cooper's hawks on his plantation actually fly low over bushes and beat them with their wings to flush out quail on which they feed. The Cooper's hawk has a long *rounded* tail that showed clearly in flight, contrasting in its shape with the square tail of the sharp-shinned hawk that was passing over.

More Cooper's hawks and many sharp-shins were coming. Suddenly, beyond them, moving swiftly toward us, we saw a dark bird that brought a shout from all who recognized it. It was a male peregrine, smaller than The Princess. He was coming at tremendous speed, quickly flapping his sharp-pointed wings. He passed the small, slower-moving sharp-shins and Cooper's hawks. Then, right above the Lookout rocks, he mounted straight up into the sky, turned over on his side and plunged down the mountainside. Halfway down the slope he darted into a column of monarch butterflies we had not noticed, which were also migrating far below us. Four times the peregrine stooped at the butterflies before he caught one with his feet. Then, while in full flight, he held the butterfly to his bill and ate it as he swept on his way southward along the mountain.

The peregrine had just disappeared when a pale gray marsh hawk, a male (the female is larger and is brownish), appeared far down the mountain slope. He was flapping and gliding lightly and buoyantly just above the trees. This long-winged bird, with a small, owlish face, had a prominent patch of white feathers at the base of his long tail.

It seemed strange to see a marsh hawk flying high along this mountain ridge. I was used to seeing them twisting and turning a few feet above the ground as they quartered low over marshes and prairies. Always they alternately flapped and glided in their low-level search for mice, frogs, or small birds, which they pounce upon with astonishing speed.

As the marsh hawk glided below us, we could look down on its long, narrow wings, with the black-tipped primary or flight feathers spread like the fingers of one's hand. I was struck by the way the marsh hawk held its wings uptilted like a turkey vulture as it flew. But I was even more impressed by the resemblance of its wings, slender body, and long tail in outline to that of a turkey vulture. Those soaring type wings, like those of the vulture's, give its light-weight body high-lift efficiency in the air. D. B. O. Savile, who has studied the wing shapes of birds in relation to their flight habits, has classified the marsh hawk with the group he calls the *Slotted-Soaring*, or *High-Lift* Wing. The group includes the eagles, osprey, the vultures, condors, and some of the larger hawks—the red-tailed, red-shouldered, and their many relatives.

The long tail of the marsh hawk, which it uses like a rudder, gives it quick turning ability in the air as it twists about in search of its prey. When its sharp eyes see a mouse in the grass, it can stop instantly in the air by quickly flapping its wings horizontally and by dropping its body and long tail downward to brake its flight. Then it stretches its long legs into the grass to snatch up the victim in its talons. Without stopping, it rises to continue its flight.

One day over a field in North Carolina, I saw two crows hotly chasing a marsh hawk. The marsh hawk was carrying a mouse in its talons. When the leading crow overtook the marsh hawk and was about to dive down on it, the hawk suddenly dropped the mouse. The crow, without checking its flight, deftly snatched the mouse out of the air and flew away. Birds sometimes practice piracy just as men do.

While I gazed after the disappearing marsh hawk, I heard a shout from my companions. I turned. Coming swiftly toward us was a large white and black osprey, or fish hawk. It was about 150 feet above our heads and was followed closely by a big bald eagle. Above the eagle, a jet black turkey vulture spiraled in slow lazy circles.

The osprey was all white below with a characteristic black "wrist" mark at the bend of each wing. Its wings, fully five or six feet across, were spread wide as it sailed swiftly by. The primaries were spread, and there was an odd crook in each wing that distinguishes the osprey in flight from all other birds.

The massive, black-bodied, white-headed eagle, an adult (the immatures or young birds are all black except a whitish area on the undersurface of the wings), was gliding along about 50 feet higher than the osprey and about 100 feet in back of it. Its broad, black, seven-foot wings were spread flat as a board, a characteristic of eagles in their gliding flight.

One of the college students near me whistled softly. He, too, was watching the eagle through his binocular. We could see its snow-white tail spread slightly, and the great yellow feet tucked straight back under its broad white tail as an airplane might retract its landing gear after taking flight. In this way a bird helps to cut down its "drag," or the resistance of its body to the air. The fierce yellow eyes of the eagle stared intently ahead, and the enormous, hooked yellow bill cut the air like the sharp prow of a ship.

The eagle, following the osprey, brought a swift memory of a clash between two of them I had seen only a few years before at Cape May, New Jersey. One August day, while bird watching among the sand dunes along Delaware Bay, I saw an osprey flapping by over my head. It glided out over the sunlit waters of the bay. Suddenly it stopped in mid-air and began to hover in one spot, beating its wings rapidly, its body horizontal, its legs and feet dangling toward the water. Its keen eyes, able to see a fish under water from a height of 100 feet or more in the air, scanned the waters below. Then it plunged down, feet first, with its wings half closed, its breast and feet striking the water with a loud splash. For a moment it disappeared except for the tips of its wings showing above the water. Then it reappeared and struggled up with a tremendous flapping and thrashing of its wings. As it shook the water from its feathers, it was quickly airborne and I saw that it carried in its feet a large fish. It held the fish, with the fish's head pointed into the direction of its flight. Apparently the osprey instinctively knew that by holding it in this way, the fish would offer less resistance to the air.

At that moment a bald eagle rushed down out of the sky. I had not seen the eagle, and it may have been watching the osprey from high above as eagles often do. The osprey is an expert fisherman; the bald eagle, a fish-eater, too, is not. As the eagle flashed downward in its dive on the osprey, the osprey struggled to fly even faster. But the eagle was too swift for it. As the eagle seemed about to seize the osprey from behind, the osprey dropped the fish. Changing its course slightly, the eagle dived downward and caught the fish neatly out of the air. Then it flapped heavily away.

For a few moments there had been a lull at the top of Hawk Mountain in the flow of migrating birds of prey. Below us, along the mountain slope, we saw blue jays flapping by (they are largely woodland birds and are not strong or swift fliers). A file of migrating crows also passed. When a flock of barn swallows came darting by, we realized that many kinds of birds follow the glider highway, whether they are soaring birds or those that depend largely on flapping flight.

Then, high over the mountain ridge, crossing at right angles to it, we saw a thrilling sight. A flock of dark Canada geese, in a long wavering V, stretched across the blue sky. Faintly we heard the wild chorus of their voices, like the distant music of baying hounds. The geese were at a tremendous height above the mountain and were not depending on the uplifted drafts of air from the ridge as they flew on a direct course southward. Each bird was flapping its wings slowly in a measured beat as it seemed to follow slightly behind the bird ahead of it. We were seeing a splendid example of formation flying, an example of ancient wild goose wisdom, known to them and a few other kinds of birds that make long flights together.

The wavering line of geese—I counted fifty in the flock—were moving far off to our right. Soon they were specks in the sky, then they disappeared in the distance. Now there was a lull in the bird flights. It was noon, a time when many birds stop their flights past Hawk Mountain and possibly come down to the mountain ridge to rest. We had not seen a golden eagle all morning. There had been a good flight of the sharp-shinned and Cooper's hawks, and many vultures. Strangely we had not seen one of the big red-tailed hawks that sweep by so commonly at Hawk Mountain.

We decided to take advantage of the lull and to eat our lunches settled down among the rocks out of the brisk north wind. Perhaps the afternoon flight would bring us other birds we had hoped to see.

CHAPTER VI

EAGLES OVER HAWK MOUNTAIN

A few turkey vultures continued to glide by during the noon hour, but another large flight of hawks did not appear until 1:15. By that time, about twenty-five people from distant places had joined us on the Lookout rocks and were watching the sky to the northeast.

A shout went up as about a dozen specks came over the broad mountaintop. Then more and more dots appeared until the sky was filled with them. They were hawks, and some of them were spiraling upward, or circling. They were too far away for us to identify them, even though we were watching through powerful binoculars.

The first birds to come within recognizable view were two red-tailed hawks. They were speeding directly toward us. Their wings, about five feet across when fully spread, were slightly retracted. As they came over about 200 feet above us, their whitish breasts shone, and we noted the contrast of the streaked feathers that extend across the belly like a dark band. This is a distinguishing mark of an adult red-tailed hawk when it is directly overhead.

At that moment, the wind was striking the ridge in great gusts. It must have sent especially powerful currents of air shooting skyward because both hawks were gliding by on "trimmed sails." Ordinarily the red-tailed hawk and other big soaring birds, when riding upward on thermals or on vertically deflected currents of air, spread their wings wide. By spreading the slotted wings, with the primary feathers opened fully, they get every possible bit of lift. But with the powerful updrafts from the mountainside to ride upon, these two had folded their wings like partly closed fans. This lessened the drag, or resistance of the air to their wings as they sped by at about thirty-five miles an hour, and it decreased their lifting

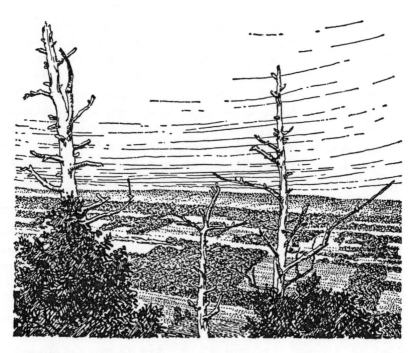

6. *A red-tailed hawk with its wings retracted.*

surfaces. By so doing, they did not have to change the "angle of attack" of their wings to keep their lift constant at their increased speed as an airplane would. We were seeing an example of controlled efficient gliding flight that since then I have often seen practiced by both the turkey vulture and the southern black vulture.

Now a large group of red-tailed hawks was approaching. Quickly we counted forty of them. They were strung out, from the farthest ridge all the way to our Lookout. Suddenly they were upon us, going over singly or in pairs. Some were rushing by, others spiraled up lazily to a greater height to circle and look down at the white faces turned up to them from below.

One red-tailed hawk mounted higher and higher over the Lookout. As it turned on its broad widespread wings to gain height, its fanned-out russet tail glowed in the sunlight. Most hawks and eagles are silent when passing Hawk Mountain in migration. But the red-tailed hawks often give a long-drawn squeal, like the p-s-s-s-s-r-r-r-r sound of escaping steam. One can imitate the sound by whistling the *p*, *s's*, and *r's* through clenched teeth. The red-tailed utters this note near its nest when it is frightened or angry. Then it may dive at the intruder, whether it is a man or another hawk.

One day, the nesting pair of red-tailed hawks that I watch each summer on a North Carolina farm were circling slowly over an oak woods not far from their nest. Suddenly the male, a little smaller than the female, as are the males of all hawks and eagles, left the woods and glided out over a cornfield. There, like a hunting dog discovering a fresh scent, he turned quickly in a tight circle and began to spiral rapidly upward over the field. He had found a thermal.

Some ornithologists who have studied bird flight believe that hawks, vultures, and eagles can in some unknown way sense the locations of these invisible rising currents of air. Certainly they move straight toward them, as though they knew their location exactly. They "get aboard" much as one might step into a helicopter to be carried upward.

The female quickly followed the male and joined him, climbing on the rising column of air just below her mate. When they were about 500 feet above the field, two turkey vultures tried to enter the column of air at the point where the male red-tailed hawk was

43

spiraling upward. He flexed his wings slightly at the vultures. It was obviously a warning. One of the vultures flexed its wings in a quick responsive gesture, but both turned and swiftly glided away. Apparently the hawk resented their intrusion on *his* column of air; also on his nesting territory.

Pilots of gliders, or sailplanes, have long known about the exceptional ability of the red-tailed hawk in seeking out rising thermals, or the orographic currents of air at Hawk Mountain. Some sailplane pilots find thermals by watching the actions of circling vultures, hawks, and other soaring birds. They claim that the thermals, or vertical "winds," are as common as the horizontal winds that help speed sailplanes and birds toward their destinations.

One day, a British sailplane pilot, Harald Penrose, while riding a thermal over England, saw a large flock of jackdaws and rooks (crows), stream excitedly up from a valley below. They flapped hard until they reached a point over a steep hillside, then spread their wings and began to spiral upward on a rising column of air. Penrose followed them and in a few seconds had entered the column of air about 300 feet above the circling birds. Round and round, up and up, his sailplane arose and under him followed the rising group of rooks and jackdaws.

Looking down on them, Penrose could study every detail of their upward spiraling flight—the swept-forward wings, with their opened or slotted tips, the fanned-out tail feathers, the gaping bills as he heard the birds cry out in what seemed joy.

Not a flap was given [wrote Penrose]. But the wings made a multitude of minute controlling movements. Nor did each bird fly at the same speed or climb at the same rate. There was a constant inter-threading and overtaking, but always roughly the same circle and the same direction of movement.

Just below altitude 2000 feet, the column of rising air ceased. Penrose left it to glide over the countryside far below. Looking back, he saw the rising birds reach the top of the air column, then one by one peel off and plunge earthward like falling stones, some tumbling wildly until they reached the woods from which they had started.

In 1937, Lewin B. Barringer, an expert American glider pilot, decided to follow the Kittatinny Ridge, the migration route of the birds that pass Hawk Mountain each autumn. He started at Ellenville, New York, and guided his high-performance sailplane on the updrafts from the ridge for 160 miles. His trip ended at Harrisburg, Pennsylvania, southwest of Hawk Mountain.

Glider pilots in the eastern United States have discovered that, usually, when their sailplanes enter a thermal occupied by a soaring red-tailed hawk, the hawk will leave. But the hawk will often react differently if the glider pilot is within the hawk's home territory.

One June morning, David Weller of the Schweizer Soaring School at Elmira, New York, was gliding in his sailplane a little under 2000 feet elevation. He was south of the airport over a wooded ridge in which a pair of red-tailed hawks had their nest. When the sailplane appeared, one of the birds was spiraling about high above the plane. Suddenly it drew in its wings and dived at the glider. Weller said that when the hawk started its dive it was about 1500 feet away. It came at him head on and passed within two feet of the canopy of his plane. As it passed, Weller said that he heard the hawk screaming. When it came out of its dive, the bird flew along just above and behind the wing of the glider as though to harass it, or to escort it out of its territory. It made several attacks on the glider and ceased them only when the sailplane had moved out of the hawk's home range.

Not one of the red-tailed hawks going over our Lookout on Hawk Mountain that day was acting aggressively. Each minded its own business and made no diving attacks on another. The forty red-tailed hawks had barely passed when a great shout went up from the watchers. A large dark bird and an adult bald eagle, flying side by side, were approaching. As they drew near we could see that they were about the same size. They were gliding just below the ridge top, and as they approached, we could look down on their backs.

Now we saw the smaller head of the large dark bird in contrast with the heavier head and larger bill of the white-headed eagle. As the dark bird drew nearer we saw the shower of golden hackle

45

feathers around its crown and down the back of its neck. It was a golden eagle—the first one most of us had ever seen!

For a moment I forgot the bald eagle in my concentration on the golden eagle. With a wingspread of about six and a half or seven feet, and weighing from nine to twelve pounds, this magnificent eagle is known around its world-wide range as the King of Birds. It is considered by men who know it well to be even nobler than the splendid bald eagle, our national bird. But the bald eagle is strictly an American; the golden eagle is not.

When falconry flourished in Europe in the Middle Ages, only kings flew the golden eagle. Chinese falconers have trained it to pursue and attack wild wolves in the mountains of Tibet. Its hunting, like that of the peregrine, is clean, and spirited. It stoops in tremendous dives estimated at speeds of 100 miles an hour or more to pick up jack rabbits from the plains of the West, or to sweep any flying birds out of the air, from grouse, ducks, geese, and swans to the large great blue heron. It has been known, in times of food shortage, to attack full-grown pronghorns and deer.

Although the golden eagle nests from Labrador southward sparingly in the Appalachian Mountains, its nesting area in the United States is largely west of the ninety-ninth meridian, in the mountains of the West south into Texas and west to California and Oregon. Its main food is rabbits and rodents and other small animals that compete with livestock for range grasses. When deprived of these animals on rangeland overgrazed by sheep and goats, the eagles may kill an occasional lamb or kid.

In the Davis Mountains of western Texas, the sheep ranchers have condemned the golden eagle for killing lambs, and claim that the bird is an economic threat to their sheep raising. There the ranchers have hired skilled aviators, flying small highly maneuverable monoplanes, to shoot the eagles out of the air. Using a sawed-off shotgun, one of these aviators reported killing 1875 golden eagles within two years. In the opinion of a Texas naturalist, who sympathized with the eagles and investigated the killing, this is an unequaled record of eagle slaughter. In the early 1940s, an organization of sheep ranchers called the Big Bend Eagle Club of West Texas, hired a pilot to shoot eagles from aircraft. Through six winters—

from 1941–42 to 1946–47—he killed 4818 golden eagles or an average of more than 800 a year.

Many years ago, as early as 1915, when the eagles were first confronted with a small, noisy plane invading their territories, some of them began suicidal attacks on it by dashing themselves against the invading monster. These self-destructive attacks by an occasional golden eagle were thought to be territorial defense. Perhaps the eagles considered the airplane a rival in the air, or they were simply attacking it to protect their eggs and young.

Another eagle-hunting aviator boasted of killing more than 8000 golden eagles while hunting them over the mountain slopes of the Trans-Pecos mountain range of Texas. Eagles, hunted by armed men in planes, simply had no chance of surviving. Conservationists knew that the appalling slaughter of golden eagles, important in helping to control overpopulations of animals from mice and jack rabbits to pronghorns and deer, had to be stopped before the eagle vanished from this country. In a congressional amendment of 1962 to the Bald Eagle Act of 1940, the golden eagle was also given complete protection from hunting. However, the Secretary of the Interior may prescribe golden eagle control, presumably by shooting or trapping, when requested to do so by the governor of any state.

As the golden eagle and bald eagle passed us at Hawk Mountain that day, we could see that they appeared about the same size, but the tail of the golden eagle seemed longer and not as squared-off as that of the bald eagle.

Now they were directly opposite our Lookout. As we turned with the passing birds to watch them, a tiny high-flying sparrow hawk, smallest of our American falcons, darted down at the golden eagle. Apparently irritated by the presence of the larger birds, or possibly playing with them, the sparrow hawk rose high into the air and again stooped at the back of the golden eagle. Both eagles paid it no more attention than if it were a mosquito and kept unconcernedly on their way.

Years later, when I talked one day at the Lookout with Maurice Broun, then Director of Hawk Mountain Sanctuary, he told me that as a rule, there was little conflict between the migrating birds.

The Sanctuary had been established in 1934, by Mrs. Rosalie

47

Edge of the Emergency Conservation Committee of New York City to protect the migrating hawks and eagles from hunters. Each fall, the hunters had come to the Lookout rocks to slaughter tens of thousands of the passing hawks and eagles with a shotgun barrage.

Broun said he had often seen the little sparrow hawks harmlessly attack the big eagles. Migrating peregrines and goshawks also made occasional passes at golden eagles, but as a rule the birds moved by in harmony. However, on a day in November 1946, there was an exception.

That afternoon, Broun was lying on his back among the Lookout rocks, gazing into the zenith through his 7-power binocular (a binocular that can magnify an object seven times). He saw at a tremendous height a large dark bird and near it, a small hawk. He could not identify either bird clearly and substituted an 18-power binocular for the 7-power one. He could see then that the larger bird was a golden eagle, but he still could not identify the hawk.

The hawk was diving at the eagle and apparently harassing it. Suddenly the eagle thrust forward and rolled over on its back as effortlessly as a fly landing on a ceiling. As it turned, belly upward, it reached out with its talons and snatched the hawk out of the air. For a moment Broun could see the hawk struggling feebly. Then the eagle, with its wings drawn back at its sides, plunged earthward at tremendous speed. Still clutching its prey, it entered a densely wooded flank of the ridge. Just before it disappeared, Broun saw the wings of its victim fully stretched from its sides and the ruddy breast of the bird. It was a red-shouldered hawk.

Apparently the eagle's hunger had been far greater than its ability to tolerate the impudence and annoyance of the red-shouldered hawk's attacks. Of the 200,000 hawks, eagles, and vultures that Broun had seen in the first twelve years of his daily watch at Hawk Mountain, this was the only time he knew of a large bird of prey to strike down a smaller one.

Late on the afternoon of that first day we were at Hawk Mountain, we saw another golden eagle come over the Lookout. It was the last bird of the day, and the sun was setting when the eagle began to spiral upward over our heads. We watched it through our binoculars as it rose on its mighty wings, with the big flight feathers of its wing tips set and outspread. Higher and higher it went, growing smaller and smaller in the last light of the setting sun. Just before disappearing, it glowed for a moment, a pinpoint of throbbing light that flared like a dying ember of gold. Then it melted into the dome of darkening sky as the first faint stars appeared overhead.

CHAPTER VII

THE GREATEST GLIDER IN
THE WORLD (DYNAMIC SOARING)

Suppose we took passage from New York City on a ship heading southward past Puerto Rico and Trinidad, deep into the South Atlantic Ocean. When we reach a point on an east-west line between Rio de Janeiro on the South American coast and Walvis Bay to the westward on the South African coast, we are approaching the Tropic of Capricorn. Beyond it lie the loneliest, windiest seas on earth, where there are no large land areas, unless we planned to go on to the frozen continent of Antarctica, thousands of miles farther south.

For days we have been steaming through tropical waters. Soon we will begin to reach the cooler part of the South Atlantic where the air is fresher, the sea bluer, and the wind blows steadily out of the west. We are approaching the latitudes 40 degrees to 60 degrees South, a part of the southern oceans that extends around the world. It is the home of the great wandering albatross.

Called by seamen the "Roaring Forties" and "Furious Fifties," for their steady, often galelike winds and giant waves, these latitudes have the roughest seas in the world. It is no accident that the wandering albatross lives there. The strong westerly winds give support to its 11½-foot wingspread, the largest of any living bird except the maribou stork of Africa, a landbird whose wingspread may reach 12 feet.

To the wandering albatross—a giant, animated glider, or sailplane —the almost constantly blowing westerlies are the power with which

51

it may circle the world several times a year and stay at sea for months or even years without ever touching land.

As our ship reaches latitude 30 degrees South (each degree of latitude is about 69 land miles from north to south) we are at the northern edge of the range of the wandering albatross, and about 2000 miles south of the equator. But unlike the comparatively lifeless warm waters we have left, the cold waters here are alive with minute diatoms* (plants), the "pastures of the sea." Upon these graze billions of small crustaceans. Some of them—the Euphausians, or opossum shrimps, called by the Norwegian whalemen, "krill"—are the principal food of fishes, seabirds, several kinds of seals, and of some of the largest whales. We are in the richest waters of the world, where thick masses of tiny animals and plants (plankton) extend fathoms deep and give life to a vast population of oceanic creatures. Here the wandering albatross has the combination of steady winds and an abundant food supply. We can expect to see one at any moment.

Then, far out over the waves, a quarter of a mile away, we see a flash of white. An enormous bird, skimming low over the waves, suddenly shoots skyward above the sea. Its body, wings, and tail make a white cross as it turns. Then it plunges down into a deep trough of the sea and disappears. Farther away, it rises into the wind to 40 or 50 feet above the sea. Downward it dives again to skim for a long way over the surface. So fast does it glide, at a speed of seventy miles an hour or more, that in a few seconds it is far away and hidden as it follows a deep trough below an ocean swell. It swoops upward into view again, turns and twists, then plunges into another trough and is gone—a lone bird in a vast ocean of blue.

It is the ease of flight of the wandering albatross that makes it a never-ending wonder to all who have seen it. A heavy bird, it weighs twenty pounds or more and has a body the length of a large

---

* "The diatoms are the connecting link in the causal sequence between the energy of the sun and the oceanic world, including seabirds. Diatoms surpass in bulk or annual productiveness all other aquatic plants a thousandfold. Because of their abundance, they are the sole food of certain marine animals, the partial sustenance of many more, and the ultimate source of life for all."—Robert Cushman Murphy, *Oceanic Birds of South America*

swan, or about 4½ feet. Yet it seems to swim through the air as easily as a fish through the waters of a quiet pool. Long ago seamen named it, "the Monarch of Oceanic Birds," and "King of the Seas."

Although man has thrilled at a wandering albatross's magnificent flight, he has been equally impressed by its voracious appetite. Because it is large, the wandering albatross must seek food incessantly to keep alive, and is almost perpetually in flight in its search for food. When it settles hungrily on the water to feed on the squids and cephalopods that often swarm near the ocean's surface, it usually "sits down to its dinner," rather than skim food from the water while it is flying. It will follow ships for days to feed on the galley refuse thrown overboard and strewn in the ship's wake. There are some ornithologists and others who believe that the wandering albatross may have the keenest appetite of any bird in the world.

Dr. Robert Cushman Murphy of The American Museum of Natural History, New York City, is one of the world's greatest authorities on the seabirds of the southern oceans. He has noted that if several wandering albatrosses are following a ship, and one is shot by a member of the ship's crew, upon its falling into the water its companions will descend on it and tear it to pieces. Dr. Murphy has written that it will even attack a drowning man.

During World War I, after a naval battle off the Falkland Islands between British and German warships, members of the crew of the British cruiser *Kent* were sent out in boats to pick up German survivors of the cruiser *Nürnberg*. The Falkland Islands, just off the southern tip of South America, are in the home range of the wandering albatross. The crew reported that some of the many hungry albatrosses that had been following the ship dived down to attack the men struggling to stay afloat in the waters.

Dr. Murphy wrote of the astonishing story of a seaman who fell overboard from a sailing ship bound for Australia many years ago. The seaman was immediately attacked by a wandering albatross that swooped toward the water and lunged at him. To a man swimming in the sea, the six- to eight-inch-long beak of the albatross with its knifelike edges is a fearsome weapon. The seaman desperately reached up and grabbed the bird by its head and neck. He held it under water until the bird drowned, then used its buoyant body to keep himself afloat. Being weighted down by heavy boots and clothing, he would have drowned had he not clung to the bird. He was picked up by the crew about an hour later. They reported that this was the first albatross they had sighted in a month.

The wandering albatross is a surface feeder and will swoop down to alight near anything on the water resembling food, especially if it is white. One of the strangest items ever reported from the stomach of an albatross came from one that had been killed in a desolate part of the southern oceans by Dr. Edward A. Wilson of the British Antarctic exploring ship *Discovery*. It was "an undigested Roman Catholic tract with a portrait of Cardinal Vaughan." Why or how the albatross chose this item from the ocean's surface has never been explained.

As our ship moves southward into the Roaring Forties, another wandering albatross appears. Here the westerlies on a typical day sweep over the ocean at thirty-five to forty miles an hour. On land we would call it a strong wind, one that tosses the branches of trees and causes men to clutch their hats to keep them from blowing away. The sea rolls with the wind in long swells that are about 300 feet from crest to crest and the troughs between them are 15 to 20 feet deep. Below the pale blue, almost cloudless sky, acres of sea-green water stretch away in heaving swells to the horizon. Our southward-steaming ship tosses heavily, buffeted by the west wind and the rolling sea crashing against our starboard beam.

The wandering albatross is usually solitary, but now three appear, and half a dozen or more may sweep around us before long, moving at high speed. They have learned that a passing ship not only stirs up the ocean's surface and the squids on which they feed, but that food scraps will be thrown overboard during the day.

A wandering albatross is trailing close behind us. It is about 50 to 60 feet above the right (starboard) side of our ship, just to the windward of our broad wake of disturbed waters. It is gliding southward slightly behind and following exactly our ship's course. At close range we can see the bird making constant small movements of its wings, feet, and tail, adjusting their surfaces to keep its "sails" trim, and to retain even flying speed in an easy unhurried way.

Those magnificent wings are one of the marvels of evolution. Spanning fully eleven feet from tip to tip, and only about eight to nine inches from front to back, in the narrow wings of the bird overhead we were seeing the most superb "sails" of any living creature. However, the wingspan of one of the ancient ancestors of the wandering albatross may have been much greater. Early in this century, paleontologists reported that the sternum, or breastbone of a giant bird, presumably an albatross that had lived during the Eocene Age, 34 to 58 million years ago, had been dug up in Nigeria. They believe that this ancient bird had long narrow wings that spread fully twenty feet, or about twice the wingspan of the wandering albatross.

In its wing structure, the wandering albatross is one of the most specialized of any bird that flies. As it glides about, spending most of its life in the air, it is thought that, at times, it may even sleep on the wing. Its wings are so long that seamen who have kept it on shipboard as a pet discovered that it could not fly from the narrow deck of a ship.

Wandering albatrosses nest for a few months of the year on remote islands of the southern oceans. While sitting on their nests, they are so gentle and fearless that a man can stroke them or pick them up. To take off from land they need a nearby runway or cliff from which they can run and become airborne. With their wings spread as they run, they lift up like kites into the wind.

What is particularly astonishing about the bird floating above us is the shape and structure of its wings. Unlike a broad-winged eagle, vulture, or other soaring landbird, its narrow outer or "hand" wing is short compared with its enormously long and narrow inner wing. The short outer wing has the normal ten primary, or flight, feathers of most birds, but the long inner wing (glider part) has at least forty secondary feathers, of which most birds have six to twelve.

The bones under the feathers that form the framework of the wings are extraordinarily light. As in other birds, they are hollow and filled with air instead of marrow, yet they are very strong for their weight. We are looking at a living sailplane that is adapted to high-speed gliding and maximum lift. We can also see that the long wings have a pronounced arching, or camber, that aids in giving the bird greater buoyancy. The aspect ratio of the wandering albatross's wings—the ratio between the wingspan and the width of the wings from front to back—is extremely high. It is said to be about 18:1, or equal to that of the most efficient gliders man has ever built. But even more efficient than the best glider is the ability of the albatross to maneuver in the relatively shallow area of about 50 feet above the ocean. This has never been experienced by a man in a glider and may never be known to him other than in theory. Soon we are to see it demonstrated by the bird overhead.

Looking through our binoculars, we could see one large brown eye of the wandering albatross turned toward us in profile. Its white body, exquisitely streamlined, is shaped almost like that of a fish.

7. *Dynamic soaring—the wandering albatross.*

The large webbed feet, set far back on its body, are carried straight backward, extending a few inches beyond the white tail. This is an old bird in the full splendor of its snow-white plumage except for its jet-black primaries and outer wing edges. It was using its feet and tail as supplementary surfaces to its wings for control of its flight.

Soaring above and behind our ship, it could not, however, keep its position without occasionally renewing its source of energy for gliding flight. Although it can flap its wings, which it may need to do in a low wind to get off the water, it depends on gliding for sustained flight. To keep up with us it must find a regularly renewable source of energy. Just as an automobile or an airplane needs a renewed supply of gas, or petrol, the albatross must take time to "refuel." It gets its energy from the wind, and we were about to see a dramatic example of how it does it.

The albatross is now overtaking us rapidly. Suddenly it turns away at right angles from the ship's course and is carried by the wind across our wake. Drawing its wings toward its sides (flexing them) until from below they resemble a flattened W, it dives rapidly away from us with the wind under its tail. It is the beginning of its famous dynamic soaring flight. Down, down it plunges in a long glide toward the sea. It started from 50 to 60 feet up, where the wind is blowing at its present top speed of 40 miles an hour. With its own air speed of at least 30 miles an hour, and carried by the strong wind, the albatross plunges faster and faster toward the sea. As it dives it enters layers of air that are moving slower and slower. The closer it gets to the ocean, the more the wind is slowed by friction of the water's surface. With lessening resistance of the air to its weight, or the pull of gravity, the albatross builds its speed more and more rapidly. By the time it reaches the bottom of its dive it is traveling more than 80 miles an hour.

Suddenly, just as the albatross seems about to strike the water, it banks steeply, with one long wing dipped not far above the sea. As it turns sharply to the right to parallel the course of our ship southward, it seems to be racing us. But now it has disappeared in a low trough of the sea where it skims low over the surface, its speed undiminished. Now it rises out of the trough, still opposite our ship. It shoots upward toward us, and into the wind, as though it will climb back into the air over our ship again. And that is exactly what it does.

As the albatross rises above the waves, still at a speed of almost 70 miles an hour, a strong updraft of air from the crest of a wave gives it an additional boost that lifts it for about 15 feet. Now its speed as it climbs into the wind, instead of slowing, increases as the bird is propelled upward and forward by its own momentum. As it rises to 20 feet above the sea, its speed is still more than 70 miles an hour and it is moving *horizontally* against the wind at about 38 miles an hour. By maintaining its high angling speed at a steady rate, the albatross is using surplus energy, accumulated by its dive, to also gain height.

By the time the big bird reaches its original height of 50 to 60 feet above and slightly in back of the ship, it is still speeding at 65 to 70 miles an hour on a slant upward and moving into the wind. Now it levels off, turns southward in the ship's direction, and travels with it as before. It has regained its almost exact position relative to the ship and has not once flapped its wings. Its dynamic soaring or the use of energy (lift and forward speed) that it gained from the wind has restored it behind the ship. Now it has the momentum to continue to glide ahead and to resume its scanning of the ship's wake below it for food. We notice that when the wind is blowing and carrying them about, wandering albatrosses alight on the water more frequently to feed.

The albatross that is following our ship suddenly swoops down to the water. When close to the surface it lifts its wings high and spreads and lowers its tail. Just before alighting, it thrusts its broad webbed feet ahead and outward to check its touchdown on the surface of the water as a skater digs in his heels to check his speed on ice. Holding its partly folded wings high, it leans forward and seizes some food from the water. After gulping it, the bird turns into the wind, paddles to the crest of a wave, and on outspread wings seems to float into the air and up from the sea.

In extremely high winds that sometimes sweep the southern oceans, albatrosses cannot maintain their positions and are swept away from the ships that are steaming into or across a gale. During these storms, the albatrosses disappear and no one seems to know where they go or what they do. Some ornithologists believe they stay aloft and ride out the wind or may settle on the water in deep troughs between the giant waves until the blow is over.

Nine of the twelve species of albatrosses in the world live south of the Tropic of Capricorn. Occasionally some of those of the southern oceans have been seen in the Northern Hemisphere, but very few seem to have reached the North Atlantic Ocean in recent times. Dr. Oliver L. Austin, Jr., of the Florida State Museum has offered an interesting theory of why they do not. In his book, *Birds of the World,* he wrote:

> The windless doldrums that stretch across both the Atlantic and Pacific at the equator are effective barriers to these large gliders, and few have succeeded in crossing them.

According to Dr. Austin, three species, however, crossed the doldrums and became established in the North Pacific Ocean—the Laysan, black-footed, and the short-tailed albatrosses—and they probably did so long before the Great Ice Age when glaciers spread and retreated over the Northern Hemisphere, beginning about a million years ago, accompanied by tremendous changes in climate.

However, scientists have evidence that albatrosses also soared on North Atlantic winds during the Pliocene epoch that preceded the Ice Age. (In the United States, scientists have estimated that the Pliocene lasted from about 12 million to 1 million years ago.) During very recent times, in the last hundred years, the fossilized bones of several Pliocene albatrosses have been dug up in what is now England and in Florida.

Dr. Austin observes that, within historic times, so few albatrosses of the southern oceans have struggled into the North Atlantic that not one has become established to become a nesting bird there. James Fisher and R. M. Lockley in their book *Sea-Birds* wrote about these strays with admiration:

> Among the two dozen casual sea-bird visitors to the North Atlantic undoubtedly the most exciting are the kings of the tubenoses*—

† "Tubenoses" is a term used by some ornithologists for a certain group of seabirds—albatrosses, petrels, fulmars, stormy petrels, and diving petrels—that have paired tubes along the ridge of the bill. Through these the birds breathe air that is carried back to their nostrils and then to the lungs. Hence the name "tubenoses," or, properly, Tubinares.

the albatrosses, whose occurrences in the North-Atlantic-Arctic are really monuments not so much to the fact that from time to time the best-adapted birds make mistakes and get right out of their range, as to the extraordinary powers of endurance and flight of the world's greatest oceanic birds. Five albatrosses have strayed into the North Atlantic. . . . All breed in the southern regions of the southern hemisphere.

Of these five kinds, a wandering albatross was apparently shot just before Christmas 1909 and ended up in London hanging with some turkeys in Leadenhall Market. Another, the yellow-nosed albatross of the southern oceans, has reached the St. Lawrence River, Canada, and the coast of Maine.

Among all southern albatrosses, the one that has wandered into the North Atlantic Ocean most often has been the black-browed. It lives generally in the southern oceans from the Tropic of Capricorn to latitude 60 degrees South. Fisher and Lockley have traced nine records of its appearance in the North Atlantic. Dr. Robert Cushman Murphy wrote in his *Oceanic Birds of South America* that the black-browed albatross is a strong flier and almost peerless in the air. Even so, he believed that, when considering the records of the black-browed and wandering albatrosses in the North Atlantic, the probability that they might have been carried there on sailing ships in the nineteenth and early twentieth centuries must always be kept in mind.

Seamen, for amusement while becalmed in southern oceans, often baited a line with pork and hooked ravenously hungry albatrosses stranded on the seas in windless weather. The seamen kept some of these as pets and then released them when their ships had returned to European waters far to the north. One of the most famed of all the southern albatrosses to reach northern Europe might have gotten there in that way.

In 1860, in the Faeroe Islands in the Atlantic Ocean north of the British Isles, a black-browed albatross from the southern oceans arrived one spring. It appeared there in company with the colony of gannets, large white seabirds, that returned each spring to nest on the rocky cliffs of Mykines Holm, the westernmost islet of the Faeroes. The albatross may have reached there aboard a ship from the southern oceans or it may have been carried there by unusually strong winds. No one knows, but the people of the islands did know that it lived there among the gannets for thirty-four years and became a tradition. The islanders saw the albatross constantly with the gannets on the rock ledges, and it fed with them each day in the neighboring waters of the North Atlantic Ocean. Each autumn it left the rocks with the gannets on their southward migration and it returned with them to the Faeroes the following spring.

The island dwellers looked upon the big bird with awe, and they called it "The King of the Gannets." Then, on May 11, 1894, a visitor to the islands recognized the unusualness of the bird and shot it. It was taken to the Copenhagen Museum. Scientists, examining it, discovered that it was a female. Its skin and feathers were preserved as a scientific specimen in the museum and for the extraordinary record of a bird that had traveled thousands of miles from its home, and had chosen to live among birds not of its own kind.

Perhaps even more astonishing was another record set by this black-browed albatross. Up to about 1961, it had held a record of having been the longest-lived wild bird of any known in the world. Then, in 1961, a European oystercatcher, a kind of shorebird that had been banded as a chick at Vogelwarte, Helgoland, was trapped on its nest at the known age of thirty-four years. When last trapped in 1963, it was thirty-six years old, and thought to be the oldest banded wild bird known to the world.

CHAPTER VIII

FROM GIANTS TO FAIRIES OF
THE SKIES (HOVERING FLIGHT)

The drone of swift-beating wings came through the soft Carolina air. Above the contented hum of bees around the honeysuckle at my door, I heard an outburst of sharp, angry squeaking. Then I saw a blur in the air as something darted by with blinding speed. The blur was two tiny birds looking no larger than bumblebees. They were chasing each other, head to tail, over the lawn and flowers of my garden.

Suddenly they shot straight upward, as though drawn skyward by an invisible string. They were face to face, only inches apart. As they rose they were thrusting their slender, rapierlike bills at each other, like small fencers on defense and attack. It was a fierce duel in the sun between two male ruby-throated hummingbirds. On this sparkling April morning, they were fighting over feeding rights to nectar in the red azaleas of my garden.

I could see the small, upright bodies of the hummingbirds silhouetted darkly against the bright sky. But their wings were moving so fast they were like faint shadows at their sides. As they continued to rise, they were beating their wings at about seventy times a second, according to Crawford H. Greenawalt,* who has timed the wing-beat rate of hummingbirds, using high-speed photography. This is almost the fastest wingbeat of any bird in the world. Only another

* Mr. Greenawalt, an engineer and executive of E. I. DuPont de Nemours Company, has made specialized studies of the flight of hummingbirds. His results have appeared in the book *Hummingbirds*, published by Doubleday & Company, Inc., in 1960.

hummingbird, an even smaller one of South America, is known to beat its wings faster. Mr. Greenawalt measured its wingbeat rate at eighty times a second.

When the fighting ruby-throats suddenly dashed away in a straight line and disappeared in a nearby woods, they were traveling before a light wind and moving at least thirty miles an hour. One summer day, near Cape May, New Jersey, a ruby-throated hummingbird suddenly appeared opposite my car and kept pace with me for a short distance along the highway. I looked at my speedometer. I was driving at fifty miles an hour. A strong wind was blowing in the direction that the hummingbird and I were traveling. Undoubtedly it helped to speed the hummer along at a far greater pace than its maximum flight speed, which Mr. Greenawalt has measured at twenty-seven miles an hour.

Ruby-throats are the only hummingbirds that live regularly east of the Mississippi River. The male is about 3 to 3½ inches long. He has a wingspan of 4 inches and weighs about one tenth of an ounce, or approximately the weight of a copper penny. He is one of the smaller hummingbirds (they range from 2 to 8½ inches long) of the 320 species that live in North, Central, and South America, and the islands of the Caribbean. The tiny, 2-inch-long bee hummingbird of Cuba and the Isle of the Pines is the smallest bird in the world. Hummingbirds are strictly Americans and live only in the Western Hemisphere. Apparently the thousands of miles of water of the broad Pacific and Atlantic Oceans have been a barrier to their crossing to Europe, Asia, or Africa.

Almost everything about hummingbirds is unusual. They live over a tremendously wide area in North and South America—from the flat seacoasts to the tops of the highest mountains. One group of South American hummingbirds lives permanently between 12,000 and 15,000 feet above sea level. The mountain nights are so cold there that the hummingbirds must lower their body temperatures, like winter-sleeping woodchucks, and go into a torpor resembling hibernation to conserve enough energy to live until morning.

In general, hummingbirds live in all altitudes and climates, in tropical jungles and temperate forests, in backyards and gardens, on plains and deserts. Hummingbirds live wherever and whenever flowers bloom. With their small size and unique ability among

64

birds to hover and to fly backward, they have been able to enter a world where they have little competition from other birds for the nectar and tiny insects on which they feed—in the hearts of flowers.

With their ability to fly long distances, some hummingbirds have even reached far into the northern part of our continent. One, the rufous hummingbird, nests north to Alaska, and migrates southward about two thousand miles each fall to winter in Mexico.

Eight kinds of hummingbirds nest well within the United States —the Anna's, Allen's, calliope, Costa's, and rufous of the Pacific coast, the broad-tailed of the Rocky Mountains, the black-chinned of the Far West and Southwest, and the ruby-throat that nests from the Atlantic coast west to the Dakotas, Nebraska, Kansas, and Texas. Four—the black-chinned, calliope, ruby-throated, and rufous hummingbirds—nest also in Canada.

Because of their small size and relatively swift flight, humming-birds are rarely caught by hawks or other predatory birds. In their elusiveness on the wing, they not only hover and fly backward, but can fly straight up and down. Their wing action is unique among all birds. They are living helicopters. Although they move their wings backward and forward, more in a figure eight than in the circular whirl of the helicopter rotor, their maneuverability is much the same. Other birds are limited to flying forward (unless carried backward by the wind), and get most of their driving power and lift from the downbeat of their wings. Hummingbirds can get power from *both* the downbeat and upbeat of their wings. I was about to see it demonstrated.

One of the fighting male ruby-throated hummingbirds had re-turned to my garden. Apparently he had chased away his rival and had come back to feed. As he hovered motionless in the air before the red flowers of my azalea bush, the greenish iridescent feathers of his back glittered in the sun. When he turned slightly toward me, I saw his immaculate white breast and underparts, and a flash of ruby fire from the dark patch of iridescent feathers on his throat. Only the male ruby-throat has this throat patch (the female's throat is white, but she has the iridescent green feathers on her back).

The throat patch of hummingbirds is called the gorget, and it is no bigger than a dime on the ruby-throat. The male uses it in a threat-display against other males and to attract his mate. He flashes it before her, while swinging like a pendulum in front of her during his spectacular courtship flights. Ordinarily the gorget of the ruby-throat appears black, but in bright sunlight, it may glow from copper and orange to ruby-red. The scattering of light passing through the special structure of these feathers causes them to change color.

In some South American hummingbirds, the iridescent feathers glow all colors of the rainbow, but it was a male ruby-throat hovering before a wildflower that John James Audubon saw in wilderness America when he described, "this glittering fragment of a rainbow . . . flitting from one flower to another, with motions as graceful as they are light and airy . . ."

Reaching his head forward, the male ruby-throat thrust his slender bill into the flower while hovering before it. Shaped by adaptive use and millions of years of evolution, a hummingbird's bill can probe deep into the hearts of funnel-shaped flowers for the nectar and tiny insects on which a hummingbird thrives. The nectar-sugar (in some flowers it is as much as 87 per cent) gives him quick energy, and the protein of the small insects he catches in the same flowers, helps his growth.

Occasionally, hummingbirds, in their hunger, eagerly slash open the base of the flower instead of thrusting the bill deep into the cup. In response to threats of this kind, some flowers have evolved devices that protect them from a hummingbird's sharp bill. The nectar is stored in special spurs, separate from the main flower tube.

A hummingbird's tongue is so long that, fully extended it reaches well beyond the tip of its bill. The front part of the tongue is split, or divided, into two rolled, muscular halves which gives the tongue the function of a double tube. Also, the outer edges are frayed or brushlike. With such an efficient tongue, a hummer can not only eat nectar, but can gather insects from flowers.

With bills adapted to probing the depths of the deepest-throated blossoms, some hummingbirds have the longest ones, relative to their size, of any birds in the world. Although the ruby-throat's is about one fifth of its 3½-inch length (more than twice as long as the bill of the similar-in-body-size, golden-crowned kinglet), the sword-billed hummingbird of Venezuela, Colombia, Ecuador, and Peru, has a 5-inch bill. It is the longest, relative to the bird's size, of any bird in the world, and is much longer than its head and body length combined. If a man had to feed as the sword-billed does, and at a comparable distance away from his food, he would need a bill and tongue that could reach at least ten or twelve feet away.

Suddenly the male ruby-throat I was watching tilted his wings to the right and swung his body sideways in a pendulumlike motion. With a jerk, he moved before another flower only inches away. There he poised, his body slanted upward at an angle of about 45 degrees, his bill pointed into the flower. He was almost motionless in the air, except for his trembling, rapidly beating wings. I was looking at him through a powerful binocular that magnifies ten times. Even though the magnification brought him so close I felt I could touch him, I still could not see the exact action of his wings they were moving so fast. But in order to hover, I knew the strange and exquisite contortions his wings had to make to hold him there.

In 1936, Dr. Harold E. Edgerton of the Massachusetts Institute of Technology, took his first high-speed motion pictures of a ruby-throated hummingbird in flight. Using intermittent flashes in a low-

8. *Note how the hummingbird's wings turn upside down and beat the air both on upstroke and downstroke, permitting it to hover.*

pressure tube in which the flashes came at 1/100,000 of a second, he took 540 pictures a second. Later, when projected in slow motion on a screen, the film revealed the hummingbird's method of flight.

While hovering and beating his wings backward and forward in a figure-eight, the ruby-throat turns his wings completely over, or upside down, on the backstroke—something that no other bird can do. This permits the forepart of his wings to cut the air on both the backstroke and the forestroke. The offsetting lift from both strokes holds him in the air against the downward pull of gravity. And the backward and forward motions of his wings cancel any tendency for him to move backward or forward. Hanging in the air, he reaches into each frail flower, without disturbing it, and gets his food in a way that no other bird can.

The ruby-throat I was watching apparently had gotten all the nectar or insects he could find in the one flower. Now he lifted his bill slightly, and with a quick jerk stood straight up in the air like an erect clothespin. It was all done in a split second. Then he backed slowly and erectly away from the flower, his wings droning like those of some large bee.

To fly backward, the ruby-throat was beating his wings forward, somewhat as a swimmer sweeps his arms forward against the water to back up. But in order to hold himself aloft against the pull of gravity, on the backstroke he was again turning his wings completely over, just as he did in hovering. The lift from the backstroke kept his small body at an even height above the ground as he backed away.

Now he tilted his wings above his head and shot upward to hover in front of a flower slightly higher on the bush. As quickly he turned his wings downward and jerked lower to hang in front of another flower. And so he circled the bush, now up, now down, feeding at each flower, then moving to another. Always, like the directional tilting of the whirling blades of a helicopter, the tilted direction of his wings pointed the direction of his flight.

When he suddenly left the azalea bush and darted away into the woods, he flew in a straight-line, horizontal flight, in the way that other birds do. He beat his wings almost straight up and down, but with a wingbeat much faster than that of other birds.

Wingbeat rate, or the speed with which a bird beats its wings, is usually timed by scientists in beats per second. Wingbeat rate seems to be geared, in general, to a bird's size. The smaller the bird, the faster its wingbeat, especially among hummingbirds. Because wing length is shorter in small birds, like a short-legged boy who takes shorter, quicker steps than a long-legged man, small birds beat their wings faster in flight than larger birds with longer wings.†

The large, nineteen-inch-long common crow is many times the size of a ruby-throated hummingbird. The crow beats its wings, in normal flight, about twice a second; the smaller domestic pigeon, three times a second; the still smaller mockingbird, fourteen times; the tiny chickadee, twenty-seven beats; the small female ruby-throated hummingbird, about fifty times a second; and the even smaller male, seventy beats a second.

Although most birds have a regular wingbeat and flight speed at which they normally fly, they can speed it up when necessary. Threatened with danger, or chased by a bird of prey, a bird can temporarily beat its wings more rapidly and can fly much faster than it ordinarily does. A British ornithologist measuring the flight speed of birds in India, discovered that Indian crows had a cruising speed of about twenty-five miles an hour when near the safety of trees where they could dive for refuge. But when flying across open country where they were vulnerable to attacks by hawks and eagles, they speeded up to thirty or thirty-five miles an hour.

Where does the tiny hummingbird get its extraordinary strength to keep its wings beating so rapidly? It does not soar or glide as other birds do, but must beat its wings constantly while it is in the air. One summer day, through an accident to one, I found out. In a neighbor's greenhouse, I found a male ruby-throated hummingbird that had died while trying to escape. The little bird had struck the glass, and the stunning impact had broken its neck. Many hummingbirds are killed each year by striking picture windows. Ap-

† Despite the large number of wingbeats per second of some hummingbirds, Crawford H. Greenawalt has come to a surprising conclusion about them when comparing them with other birds. *In proportion of the length of their wings to body weight*, hummingbirds beat their wings less rapidly than ordinary birds. Mr. Greenawalt came to this paradoxical conclusion, based on his aerodynamic studies of birds, and the fact that ordinary birds generate power of flight only on the downbeat of their wings; hummingbirds get propulsion and lift from both the downbeat and upbeat of their wings.

parently they fly into the glass because they see the lawns and trees reflected there as an extension of the landscape.

When I picked up the exquisitely colored bird, I felt a great pity for it. But I also wanted to preserve its beauty.

Inside my study, I carefully removed its skin with a scalpel and scissors, a process I had learned when I worked in the Bird Department of a great natural history museum. The tiny feet were as delicate as those of a woodland shrew, but sufficiently strong to enable it to cling to a twig or other perch. When I came to the bird's wings, I examined them carefully, for the mystery of its strange and unusual flight lay in their shape and their rigid bone structure.

The wings were small and pointed, with slightly rounded tips, and they swept back to the bird's sides like the wings of a speedy aircraft. It was a typical high-speed wing, like that of my falcon, The Princess, and that of the plovers and sandpipers I see migrating each autumn along the seacoast, and that of the chimney swifts and swallows that dart about so quickly in the summer skies. But what made its wings different from all other birds was its rotating movement directly from the shoulder.

All birds, except hummingbirds, move their wings at three places—the shoulder, elbow, and wrist. They really fly from the wrist out, on the outer wing with its long flight feathers—the so-called "hand wing" or propeller. While flying a bird uses the inner part of its wings to help hold it aloft, as the rigid wings of an airplane provide the lift to keep it in the sky.

Not so the hummingbird. The entire wing is "hand wing" or propeller, which is why a hummingbird does not soar or glide. It moves its wings completely from its shoulders, which gives the wings their astonishingly free movement and the hummer's maneuverability in the air.

But the power behind its rapidly beating wings is in the hummingbird's breast muscles (called the pectorals). These make up the true motor of all flapping birds. The flight muscles of the strongest fliers are about 15 to 25 per cent of the bird's total weight. The relatively enormous breast muscles of the ruby-throated hummingbird are about 30 per cent of its weight.

The flight muscles of birds are of two colors, or of two types

70

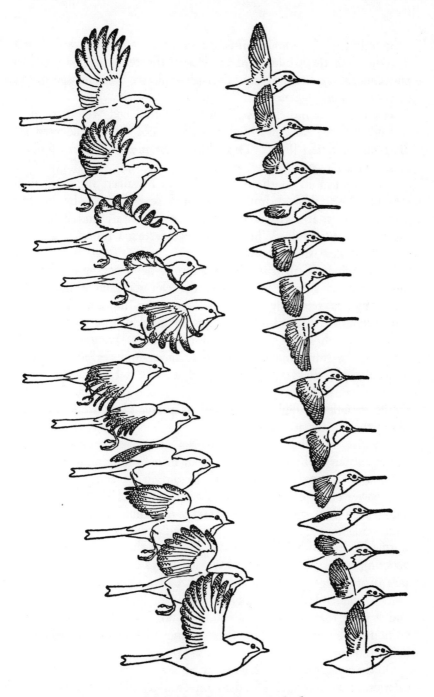

9. *The chickadee in flight (left) and the hummingbird.*

—the red ones of hummingbirds, ducks, geese, and other swift-flying migrants, and the white flight muscles of the resident quail, grouse, pheasants, and other birds that fly with explosive bursts of speed, but over only short distances.

The red flight muscles, well supplied with blood vessels, oxygen, and other fuels, have much greater endurance and are adapted for the long migration flights. The white flight muscles, short of blood vessels and oxygen, are not capable of long, sustained flight. One day, on a farm in southern New Jersey, I walked up to a quail that had been flushed from the ground several times by hunters and their dogs. It was exhausted. I caught it easily and held it in my hands for a while. When it was well rested, I tossed it into the air, and it flew strongly away.

In the wings of all flying birds, a strong depressor muscle powers the downbeat. But in most birds, the elevator muscle that *raises* the wings is relatively weak. Not so in a hummingbird. The elevator muscle that raises its wings so quickly and powerfully is relatively much larger than that of other birds. Here was another secret of the hummingbird's ability to hover and perform its other unusual flight tricks when moving about like a helicopter.

I still had to learn how a hummingbird can get so much energy for its remarkable flight. How does its power plant work—the fuel plant—that drives its motor?

Scientists use the word "metabolism" to measure or define the energy output of living creatures, from man to hummingbirds. Basically, metabolism in living things is said to be equivalent to the horsepower of an engine, or the kilowatt rating of an electric stove. Those who have studied the physiology of hummingbirds claim that these birds have the highest energy output per unit of weight of any warm-blooded animal in the world. The ruby-throated hummingbird hovering before the azaleas in my garden has an energy output per unit of its weight that is ten times that of a man running at nine miles an hour. This is the highest output of human energy known, and a pace that a man can maintain no longer than about half an hour. A hummingbird can fly continuously for much longer than half an hour.

Ordinarily, the daily energy used by a man is about 3500 calories. A hummingbird leading an ordinary life of eating, flying, perching, and sleeping, if calculated to the equivalent of a 170-pound man, would use about 155,000 calories a day. If we convert this, for comparison, into the amount of food required by a man to use the energy output required by a hummingbird, the result is almost unbelievable.

A man normally will eat about two to two and one-half pounds of food a day. If he used energy each day comparable to that used by a hummingbird, he would need to eat 285 pounds of hamburger, 370 pounds of boiled potatoes, or 130 pounds of bread.

Sugar is the principal food of hummingbirds. It has a much higher potential energy content than the meat, potatoes, or bread that people ordinarily eat. Each day, a hummingbird eats about half its weight of sugar. This is an extraordinary amount for such a small bird, but it must do so to maintain its energy output or to "refuel its engine."

The ruby-throated hummingbirds in my garden, in spring and summer, feed heavily and regularly to keep up their energy supply. They come every day to whatever flowers they like that are seasonally in bloom—to azaleas, honeysuckles, columbines, morning glories, hollyhocks, cardinal climbers, spider plants (Cleome), garden lilies, and many other "hummingbird flowers" that I have planted for them. They also hover before or sit on the perches of the small vials of sugar water that I have hung on wires suspended from trees or shrubs in the garden.

One year, between May 1 and August 15, they drank about ten pounds of sugar mixed with water that I had put in the eight small feeders. I mix it one part sugar to nine parts of water, as a rich mixture can be harmful to the livers of hummingbirds.

In spring, to attract hummingbirds that are new to my garden, I wrap red ribbon around the vials, or daub their spouts with red fingernail polish, because all hummingbirds are strongly attracted to red.

For a long time I wondered how the tiny ruby-throated hummingbirds of my garden could possibly migrate across 500 miles of open water of the Gulf of Mexico. Yet, they have been seen by observers from ships far out in the Gulf, both in fall and spring migration. Some of them flew about twenty-five feet above the water and kept strongly on their way. They made no attempt to alight and rest on the ships as some birds migrating across the Gulf will do.

In 1953, a physiologist published a report of his careful tests of the energy expended by captive hummingbirds in hovering flight. He measured their energy used while they hovered in a bell jar. According to his calculations, a ruby-throated hummingbird could not possibly cross the Gulf of Mexico. It would not have enough body fat in reserve to carry it that far, without refueling. Yet, as one well-known ornithologist humorously remarked, apparently the hummingbirds had not read the report because they continued to cross in fall and in spring, year after year. What was misleading about the physiologist's calculations?

In October 1961, two ornithologists, both friends of mine, published a report that seemed to explain the puzzle satisfactorily. There was nothing wrong with the physiologist's calculations. He had based them, however, on a hummingbird's *hovering* flight, which, according to my friends, uses more energy than straight flight. In their investigation in Georgia and in Florida, they discovered that ruby-throated hummingbirds killed in fall by striking a TV tower along the Gulf coast, were extremely fat (more than 50 per cent of their usual weight) just before they were to take off on their migration flight over the water.

Using different methods of calculating, they maintained that ruby-throated hummingbirds not only can cross the 500 to 600 miles of water, but had enough reserves of body fat (fuel) to travel about 1500 miles. This, they said, would carry them far enough to start their migration far inland in Florida and Georgia, and to end their flight well into Mexico.

CHAPTER IX

HOW HIGH DO BIRDS FLY?

One spring day in 1945, not far from Galveston Bay in Texas, Captain Neil T. McMillan of Eastern Airlines was at the controls of his commercial airliner. He was flying in clear skies above a heavy cloud overcast at 3700 feet. Suddenly his plane struck a small bird. Captain McMillan was much interested in birds and he often looked for them during his flights. He was skilled at identifying them in the air, too (he once saw swallows flying at 8000 feet).

When McMillan landed, he discovered the feathered remains of a catbird on the wing of his plane. He realized that this was an unusual height at which to see a small bird—especially one that is not usually a high or a swift flier, and spends most of its time near the ground.

The catbird, along with cardinals, creepers, thrashers, sparrows, finches, chickadees, warblers, thrushes, and other songbirds, has the so-called "elliptical" wing shape. It is a short, blunt pliable wing. Apparently it has developed widely among those birds that dart about through small openings in brush and woodlands. It is a wing shape easily seen in the familiar English, or house, sparrow of our gardens and tree-lined city and country streets.

The wing's shortness and broadness makes it highly efficient, not as the long, narrow flat wing of the swallows and falcons does for their long swift flights, but for short flights at moderate or slow speeds

Often deep-cupped, or arched, it is a wing that gives most medium-sized and small birds quick, high lift and good flight control. It is a wing shape shared by some other unrelated birds that live in or around woodlands and thickets—grouse, pheasants, wild turkeys,

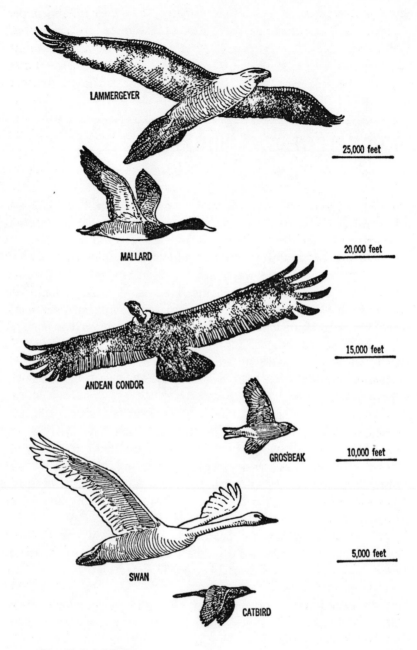

LAMMERGEYER

25,000 feet

MALLARD

20,000 feet

15,000 feet

ANDEAN CONDOR

GROS'BEAK

10,000 feet

SWAN

5,000 feet

CATBIRD

10. *How high birds fly.*

quail, woodpeckers, doves and wild pigeons, the woodcock and snipe. Unlike the long broad wings, square or round-tipped, of the eagles, hawks, and vultures that soar slowly at great heights, its short blunt shape is also quite unlike the long tapered wings of the sometimes high-gliding and high-soaring gull. That is why it seemed unusual for Captain McMillan to meet a catbird at such a great height. What had caused it to fly so high?

Apparently the catbird was migrating, and it was not only surprising to see it at such a height but to find it migrating by day. Possibly 90 per cent of all bird migration is believed to be at night.

At that time, it was thought that most birds migrated no higher than about 3000 feet above sea level, but it appeared from Captain McMillan's observation that the catbird had gone to unusual heights to reach the clear skies above the cloud overcast.

It was not until 1963—eighteen years after McMillan's experience —that Frank C. Bellrose and Richard R. Graber, two ornithologists in Illinois, announced the results of their studies, which suggested that the catbird's unusual height and cloud-piercing performance might be common to many small North American birds.

By using radar beams to detect birds migrating at night, they noted that many of them flew at greater heights during a cloud overcast. This suggested to them that the birds try to rise through clouds and fly above them. They reported that birds will even fly through rain showers. Other ornithologists have noted that the migrants will fly above low-lying banks of fog, but sleet or a steady rain may ground them altogether.

Very little was known of the height at which North American birds fly during night migration until 1960 and 1961. In the fall and spring of those years, at Cape Cod, Massachusetts, an ornithologist-physicist, I. C. T. Nisbet, studied the problem by using a radar height finder. It permitted observations on the radarscope of birds moving across the night sky at various heights, except those flying very low, within 600 feet of the ground.

Nisbet discovered that three or four hours after sunset, the most frequent height of migration was usually between 1500 and 2500 feet above sea level, and that most birds were flying below 5000 feet. However, on May 18, 1961, his radar tracked many birds flying over the Cape between 6000 and 9000 feet high, with some up to

15,000 feet. A few nights later, his radar picked up two bird echoes that were 19,000 feet high.

On several September nights in the fall of 1961, Nisbet got strong bird echoes that were between 8000 and 15,000 feet high and some scattered ones up to 20,000 feet. Those, from great heights, were usually most numerous after midnight but also just before and after sunrise. The high-flying birds were speeding at forty to seventy knots or about forty-six to eighty miles an hour and moving southeastward out to sea. From their speed and great height, Nisbet believed they were shorebirds migrating directly across the Atlantic toward eastern South America and the Lesser Antilles. He thought that they might be black-bellied plovers, semipalmated sandpipers, and certain other shorebirds known to migrate through the Cape Cod region in fall and spring during the particular times of his radar observations.

In earlier studies in England, David Lack, using radar, had concluded that the common passerine birds (songbirds) that visit Britain in winter, migrate mainly below 5000 feet, but tend to fly higher in spring than in autumn, and higher at night than by day. He occasionally recorded what he believed were small passerine birds flying over that were 14,000 feet above sea level. In September his radar detected some of these up to 21,000 feet.

Even though radar is helping to determine the heights at which migrating birds fly, and their behavior and direction of flight in various kinds of weather, it cannot absolutely identify which birds the radar beam is "fingering" in the darkness overhead. However, some of the British observers, Lack in particular, feel that many of the small birds flying over in the darkness can be identified by inference. For example, if the radar beam has indicated the direction of the migration flight the night before, and a large number of birds appears suddenly next morning at the nearest landfall in that direction, it can be assumed that many of these were birds that the radar beam had picked up the preceding night. But for positive identification of birds flying at great heights, we must still rely on direct observation.

One autumn night many years ago, somewhere over Montana, a DC-3 airliner, climbing steadily at 8000 feet (about a mile and a half above the earth), ran into a flock of wild swans. When the crippled plane landed, a portion of one of the struck birds taken from the wing weighed eleven pounds.

Another plane, flying at night in the western United States at the same altitude, roared into a flock of wild cranes. A plane, while crossing the Andes Mountains at 17,000 feet (about three miles up) collided with a South American condor that was soaring at that tremendous height. The condor made no attempt to get out of the path of the oncoming plane.

A similar accident many years ago ended almost tragically for Captain Harry Musick and his co-pilot, M. Gould Beard, of American Airlines. They were over Louisville, Kentucky, and flying directly toward a group of high-soaring turkey vultures. Captain Musick expected the birds to get out of the way, but they did not. The plane struck three of them, and one came crashing through the windshield. It knocked Musick unconscious. Beard took over the controls and landed the plane safely at the Louisville airport.

Although birds no longer can crash through the much stronger windshields of modern aircraft and knock the pilot unconscious, they still cause accidents. Planes striking large, heavy birds such as swans, may have the tail parts of the plane damaged; even small birds, such as starlings, sucked into the jet engines at take-off, have caused plane crashes.

Most of the world's greatest altitude records for birds have come from the high Alps and Himalayas during the spring and autumn migrations of European and Asiatic birds. Godwits and curlews—long-winged, swift-flying "wading" birds, or shorebirds—have been seen flying past Mount Everest at 20,000 feet. Storks and cranes were sighted passing over the Himalayas at 14,000 to 20,000 feet, and British Army Colonel Richard Meinertzhagen saw choughs (crowlike birds) at 21,000 feet above sea level, and a wall creeper over Karakorum at that height.

It was an airliner that provided me with the highest altitude record I have been able to find for a positively identified North American bird.

At 4:15 P.M., July 9, 1962, a Western Airlines L-188 Electra was cruising between Battle Mountain and Elko, Nevada, at an air speed of 345 knots. The plane was flying at altitude 21,000 feet when the pilot noticed a light thud. Later, passengers riding in the rear of the plane said they had "felt a small explosion."

Upon landing, a member of the crew discovered a dent the size of a football on the leading edge of one of the plane's horizontal stabilizers. A feather taken from the point of impact was sent to the U. S. Fish and Wildlife Service for identification. The bird was a mallard duck.

Possibly the highest bird altitude ever recorded came from the British-sponsored Mount Everest Expedition of 1921. Dr. A. F. R. Wollaston sighted a lammergeyer, or bearded vulture, at 25,000 feet. The lammergeyer is one of the largest birds of prey in the world, with a wingspread of more than 9 feet. There is an old record of geese photographed against the sun in India at an estimated 29,000 feet, but the estimate is not considered accurate and is unacceptable to ornithologists today.

In the Andes, another mountain-climbing ornithologist, using an aneroid barometer, measured the soaring altitude of a South American condor which has a 10-foot wingspread. It was 19,800 feet.

Most of the high-altitude records for birds sighted and identified over Europe were much lower: rooks (crowlike birds) over France and Germany up to 7500 feet; swifts over Switzerland, Holland, and France at 4000 and up to 9500 feet (almost two miles) and some at 11,000 feet, according to James Fisher, British ornithologist; ducks and geese over France at 7500 feet; and lapwings, a European plover with a remarkable courtship flight, at 6000 feet and occasionally to 8500 feet during its migrations.

During World War I, an Allied airman over northeastern France was astonished when his warplane passed a flock of linnetlike birds (somewhat like our house finch and purple finch) at 10,000 feet. At that time, it seemed an amazing height for a small songbird, even during migration. But other height records for small birds taken since that time, suggest, as radar has, that they may frequently fly at extraordinary heights while migrating.

On October 31, 1956, between 9:00 and 9:15 P.M., Mr. Francis Drake, while flying a single-engine Beech Bonanza airplane from Sacramento to San Bernardino, California, felt a dull thud in the forward part of his plane. He was flying at 10,000 feet at the time, and upon landing at San Bernardino, he found the foot, tarsus, and shank of a sparrow lodged in the air intake of his craft. Upon examination by an expert, the bird was identified unquestionably as a golden-crowned sparrow that had apparently been migrating when it was struck.

Another high-altitude record for a North American songbird, even more astonishing, was reported by Johnson A. Neff, a federal biologist from Colorado.

At 8:00 A.M. on March 18, 1964, the pilot of a Piper Apache was flying over the Front Range of the Rocky Mountains about five miles south of Boulder, Colorado. During his descent toward the Jefferson County Airport, he saw two small birds in flight. One struck the windshield, crashed through the glass, and landed in the cockpit. The pilot immediately checked his altimeter. It read 12,500 feet. Upon landing, the pilot had the remains of the bird checked by the Denver Wildlife Research Center of the U. S. Fish and Wildlife Service. The bird was a female evening grosbeak which had been banded in Denver on January 31, 1964. At the point where the two evening grosbeaks, "undoubtedly a pair," according to the report, had been flying when the collision had occurred, they were about 6400 feet above the mountains. It was the first report of a plane striking an evening grosbeak.

Harald Penrose, a British test pilot and student of birds, had a charming experience one calm August evening. While soaring in his glider plane at sunset at about 2000 feet, he turned into a great bubble of warm, rising air. Suddenly a European swift darted across his path. It circled, came back, then soared on set wings upon the same bubble of air on which Penrose floated.

For a few moments they soared together—man and bird—not fifty feet apart. Then the swift turned southward and vanished in the dusk. Penrose believed that the European swift sometimes glides all night and goes to sleep on the wing, high above the earth. In 1956, Emil Weitnauer verified this. After tracking some birds with radar in the night sky over Switzerland, he was eventually able to intercept them in a small airplane and with a powerful beam of light discovered they were swifts.

Swifts and swallows seem to spend more time in flight than almost any other landbirds. Leonard Wing, an American ornithologist, once wrote that "an ingenious calculator"—a man who had banded a chimney swift that had lived for nine years—estimated that it must have flown 1,350,000 miles during its lifetime. This included its nine round trips from the United States (its summer home) to South America, where it winters. For each additional year of its life, said the birdbander, another 150,000 miles of air travel could be added.

A European swift is estimated to fly at least 560 miles a day during its nesting season; a small, rather weak-flying European tit, similar to our American chickadee, flies about 62 miles a day while traveling back and forth to feed its nestlings.

I have often wished for a practical method of estimating closely the height at which I see birds flying. After years of experience I can make a fair guess, but only because I know the size of the bird I am watching. The farther it is from me, the smaller will be its relative size.

Years ago European ornithologists made some experiments that helped them estimate the height of a flying bird. In 1911, Friedrich Karl Lucanus, a German ornithologist, suspended life-sized images of certain European birds from a balloon. Knowing the height of the rising balloon, he could tell at what heights the birds were still distinguishable, and at what heights they disappeared from human vision.

A European sparrow hawk, about the size of its relative, our American Cooper's hawk, could be recognized at 800 feet, was a dot in the sky at 2000 feet, but could not be seen at 2800 feet, or a little more than half a mile.

A rook, a bird much like our American, or common, crow, was distinguishable at 1000 feet, was a spot at 2600 feet, and disappeared at 3300 feet.

The fan-shaped tail of a golden eagle could be recognized at 1500 feet, but was only a dot at 2000 feet. The life-sized image of the eagle itself was clearly visible as a soaring bird at 3000 feet, was still recognizable *as a bird* at 5000 feet, became a mere dot at 7750 feet, and was invisible at 8750 feet.

A three- to four-inch-long North American hummingbird on a utility wire appears as a dot to the unaided eyes when it is 100 feet away, but a five- to six-inch-long swallow is clearly recognizable at that distance and becomes a dot at 250 feet.

I have often wondered how many miles of the earth's surface a bird can oversee, looking in any direction from its vantage point at different heights in the sky. Colonel Richard Meinertzhagen, the British army colonel with vast curiosity about birds, once figured this out. On a clear day, when the bird's view is unimpeded by clouds or fog, it can see the following distances from various heights:

Height of Bird Above the Ground	Distance from Bird to Visible Horizon
500 feet	27 miles
1,000 "	39 "
2,000 "	55 "
3,000 "	67 "
4,000 "	77 "
5,000 "	86 "
10,000 "	122 "
20,000 "	173 "
26,700 "	200 "

CHAPTER X

HOW FAST DO BIRDS FLY?

Nothing seems to provoke an argument quicker among some people than, "How fast do birds fly?" The arguments seem to arise over two main points—whether or not the bird was flying with or against the wind, and how accurately, or by what means, the flight speed was measured.

One September day at Cape May, New Jersey, with several renowned American ornithologists, I was driving over a gravel road that penetrated a great salt marsh. We were in search of a southern bird that would be rare that far north—a reddish egret which had been reported on these marshes, but had not been verified by our group. Incidentally, we never saw it.

However, we did see an estimated 10,000 migrating tree, cliff, and barn swallows, darting about over the tawny autumn marshes, catching insects on the wing. In their food-hunting flights, some of these swallows have been timed accurately with an automobile speedometer at twenty-eight to thirty miles an hour. This seems an astonishing speed at which to catch small insects out of the air, yet it scarcely compares with the recorded accomplishment of some European alpine swifts.

One day, Colonel Richard Meinertzhagen, a British authority on the flight speeds of birds, was seated near the snowy summit of Mount Ida in Crete. He was watching hundreds of alpine swifts feeding all around him. The birds were flying at a speed he estimated at about eighty miles an hour, yet they were sweeping small beetles out of the air with their mouths while in full flight—insects so tiny that each was smaller then the head of a pin.

I was driving parallel to a canal by the side of the road that September day near Cape May when suddenly a kingfisher flew up from its perch at the edge of the water. With a loud, rattling call, the bluish bird flew along abreast of our car. Excitedly one of my companions urged me to measure the bird's speed. I looked at my speedometer. While the kingfisher kept pace with us for about a hundred yards, it traveled at forty-five miles an hour. Later I had my car speedometer checked and found that it was accurate to within a few miles an hour.

A wind blowing with us no doubt was helping speed the bird along, but we had no way of measuring the force of that wind. Therefore we had no means of measuring the true air speed of the kingfisher—how fast it could fly in still air without the assistance of the wind.

Ornithologists and other scientists agree that if a wind is blowing, unless you know the speed of the air mass (wind) with which the bird is being carried, or against which it is flying, there is no way of knowing its true air speed. However, what interests many people is how fast a bird gets over the ground, regardless of the force or direction of the wind. How fast do birds fly ordinarily? How much faster can they fly when pressed by danger, hunger, or fright?

According to Colonel Meinertzhagen, some birds fly much faster during their courtship flights, others when pursuing their prey, and others when being chased by a predatory bird. Among still others—in flocks of migrating shorebirds and swallows for example—each bird may fly faster when in the flock than when flying alone.

The colonel's studies also showed that most birds have a reserve of speed they can exert to carry them a third to twice their normal, or "cruising" speed. The difference between the two may be likened to the difference between a boy trotting and a boy running. Most birds, in flapping flight, ordinarily "trot" or "jog" along with a regular, little-changing wingbeat that is suited to their body size and the length and shape of their wings. They speed up only when necessary, and they do it by flapping faster.

One day, two friends of mine cooperated with me in measuring the straightaway flapping flight speed of my peregrine falcon, The Princess. She was not afraid of automobiles, as she rode with me, perched by my side on the seat of the car, almost every day. Along the edge of the pasture field where she took her daily flights, a good hard road ran in a straight line.

I stationed one friend in my car on the road near me, with the car motor running. The other I sent down the field, just off the roadside, about a thousand feet away. When I cast off The Princess toward him, he was to produce the baited lure suddenly and swing it around his head to attract her attention. If she flew toward him, it would be on a direct course, parallel to the road.

It worked beautifully. When I released The Princess, my friend down the field began to shout and swing the lure about his head. At the same time the friend in my car had already started and was able to keep pace with The Princess as she rapidly took off. She saw the lure and instantly started for it. Her direct flapping flight in the beginning was about forty miles an hour, but she quickly speeded up until she was flying at sixty miles an hour when she struck at the baited lure.

It was not a completely satisfactory test as we knew that The Princess could be stimulated to fly much faster if she were chasing a live bird. But we never had an opportunity to measure her flight speed against living quarry. However, I had seen her overtake domestic pigeons, and some of these birds are known from flight tests to be capable of flying 60, 82, and 90 miles an hour over a straight course. And at least one peregrine in her dive, or "stoop" at a wild duck had been timed by an airplane at more than 175 miles an hour, which was the air speed of the plane as the falcon passed it. The peregrine is fast, but apparently she is not the swiftest bird in the world, as I was to find out. However I have never seen an American bird that a wild peregrine could not catch if it seriously tried to do so.

Some of the swiftest American birds are the little sandpipers of our coastal beaches, and peregrines are known to catch them. In 1941, T. T. McCabe, an American ornithologist, while in flight training for World War II, had frequent opportunities to measure the flight speeds of flocks of red-backed sandpipers, godwits, and curlews. Of these birds, the sandpipers especially have the long pointed "high-speed wing" of the falcons, plovers, swifts, hummingbirds and swallows. While piloting his plane over San Francisco Bay, McCabe noted that flocks of these birds, flying over the waters below him, did not travel more than forty-five to fifty-five miles an hour. He began to doubt that they could fly faster.

Then, on the evening of April 5, 1941, just before dusk, McCabe was flying over the bay at 1500 feet altitude, at an air speed of ninety miles an hour. Suddenly two small, tight flocks of about one hundred red-backed and other small sandpipers overtook him. They crossed his flight path, just in front of the nose of the plane. From the angle of their flight and speed as they drew away from him, McCabe estimated that they were traveling not less than one hundred and ten miles an hour, and possibly "a great deal more."

After I had tested the flight speed of The Princess, I learned that Richard M. Bond, a California falconer, had tested the speed of a pigeon hawk he had trained for falconry. The pigeon hawk is a falcon, not as large as the peregrine and not as swift. However it is extremely agile in flight and has been known to feed much on dragonflies that it snatches out of the air while in full flight.

Andrew Bihun of the National Audubon Society told me that on September 16, 1965, he saw a pigeon hawk at the Upper Montclair, New Jersey, quarry, a high point where migrating birds pass in fall, snatch a swallow out of a migrating flock in a thrilling maneuver. The pigeon hawk singled out one bird in the passing flock and flew beneath it to get on its "blind" side. As it overtook the swallow the hawk turned over on its back, reached up with its talons, and plucked the swallow out of the air.

Richard Bond discovered that his trained pigeon hawk, when coming to the baited lure, flew about thirty miles an hour. But in chasing live California quail, shrikes, and meadow larks, she overtook them easily and could fly at a top speed of forty-five miles an hour. She could even catch a strong adult pigeon, if the pigeon started its flight when only about fifty feet out in front. But once the pigeon gained full speed, it easily outdistanced the hawk.

Near Richvale, California, Donald Dudley McLean chased a number of different kinds of birds along a highway with his car. He got some highly interesting speed records. One day he paced a little green heron at thirty-four miles an hour, with no wind blowing, and a barn swallow skimming low over the road ahead of him traveled at forty-six miles an hour. A cinnamon teal, a small duck, flew in leisurely flight ahead of him at thirty-two miles an hour. McLean speeded up, at which the duck also flew faster. When he overtook it, it was flying fifty-nine miles an hour just before it swung away from the road and flew across a field.

Two red-shafted flickers (a large woodpecker much like the eastern yellow-shafted flicker) flew at forty-three and forty-four miles an hour, and a savannah sparrow, a small bird of open fields, traveled at thirty-seven miles an hour; a second one, on being hard-pressed by McLean in his car, flew at forty-two miles an hour. It turned off the road when McLean speeded up to try to make it fly faster.

Many years ago, Alexander Wetmore, a noted American ornithologist, and former Secretary of the Smithsonian Institution, listed some of the general ranges of the flight speeds of birds. These were based on his own careful observations and those of others whose measurements were acceptably accurate. Some of them were:

Flight Speed in Miles Per Hour

Crow Family (Corvidae)	31 to 45
Small Perchers (larks, pipits, buntings)	20 to 37
Starlings	38 to 49
Geese	42 to 45
Ducks	44 to 59

There are exceptions to these figures—for example, a flock of pintail ducks chased by an airplane were timed at 65 miles an hour, and some old-squaw ducks near Toronto, Canada, were accurately timed over a measured course at 53.9, 61.5, and 72.5 miles an hour, with a wind behind them of 11 miles an hour.

In the summer of 1967, John T. Lokemoen reported his stopwatch measurements of the flight speeds of wood ducks over a measured course along the Wisconsin River. There was little or no wind to affect their flight. Nine flew at an average speed of 46 miles an hour one August day, and eight flew at an average speed of 48.3 miles an hour the following day. The wood ducks were not alarmed, and their motivation for flight was to reach their roosting area before dark. The slowest duck flew at 39 miles an hour; the swiftest at 55 miles an hour.

At Hawk Mountain, Pennsylvania, in the fall of 1942, Maurice

Broun and an associate got some interesting speed records of hawks and other birds migrating along the ridge. Most of the birds did some flapping as well as gliding, but Broun reported that many rarely flapped if the wind was blowing at fifteen miles an hour or more. Apparently this wind velocity created updrafts sufficiently strong that hawks, eagles, and other migrating birds could make long glides without flapping.

A turkey vulture glided steadily at thirty-four miles an hour; a goshawk passed by at thirty-eight miles an hour. Some birds showed a tremendous variation in their flight speeds. Sharp-shinned and Cooper's hawks moved by from as slowly as sixteen and twenty miles an hour to sixty miles an hour; red-tailed hawks from twenty to forty miles an hour; and golden eagles and bald eagles from twenty-eight to forty-four miles an hour. Sixteen ospreys varied from twenty up to eighty miles an hour for one bird. Marsh hawks traveled from twenty-one to thirty-eight miles an hour; sparrow hawks from twenty-two to thirty-six miles an hour. Fifteen crows ranged from seventeen to thirty-five miles an hour.

To me, one of the most surprising records of the flapping flight speed of a bird was that of a common loon. I had always known from watching loons fly in migration that they were fast, but I had not suspected their full potential. Flight speeds of fifty-three and sixty-two miles an hour had been reported for the common loon, which is a waterbird with a relatively small, primitive wing, not especially efficient for flight. However, there was no doubt of the heavy-bodied loon's swiftness once it got under way.

In March 1948, James A. Pittman, while flying a Piper Cub J-3, just north of Charlotte, North Carolina, and about three miles from the Catawba River, saw the loon cross diagonally in front of his plane. He was flying at altitude 1200 feet, and turned immediately to follow the bird, which went into a shallow dive. Although Pittman opened the throttle fully, the loon pulled steadily away. Just when he had gone so low that his plane was in danger of crashing, he pulled out of his dive; but he had gotten a record of the bird's speed. During the first part of the dive, Pittman was traveling at 90 miles an hour, but as the loon grew frightened it flew even faster. Pittman estimated its speed at between 80 and 100 miles an hour.

11. An airplane following a loon in a dive.

G. Ronald Austing, an ornithologist and falconer who has especially studied red-tailed hawks, discovered that their top speed is about 35 to 40 miles an hour, but he believes that diving earthward from a great height, as the male often does during his courtship of the female, it may fly at least 120 miles an hour. Austing has clocked numerous adult red-tailed hawks in horizontal flight with a special police speedometer in his patrol car.

Among gamebirds, a woodcock has been timed at more than

thirty-five miles an hour near Lufkin, Texas, wild turkeys at thirty-eight to forty-two miles an hour, and prairie chickens at forty-two miles an hour.

Lawrence H. Walkinshaw, an American ornithologist who wrote a book about the sandhill crane, timed these big birds in flight at twenty-five to thirty-five miles an hour.

One of the truly modern and accurate ways of measuring the flight speed of birds was tried in the summers of 1963 and 1964 by Gary D. Schnell, then a graduate student at Northern Illinois University. He borrowed a Doppler radar unit developed to measure the flight speeds of birds by a staff member of The American Museum of Natural History. It operated on the same principle as radar used by police to detect speeding violations by automobile drivers.

Working in northern Michigan, Schnell took his portable radar unit among colonies of nesting gulls, terns, and swallows. He got 1628 flight speed records of 17 species of birds—more speed records than had been recorded for birds in all history up to the time of his experiments. With his radar unit, which had an error of no more than a mile an hour, he found that herring gulls flew from twenty-four to about thirty-seven miles an hour; the common tern, twenty-five to twenty-seven miles an hour.

Other bird speeds measured by Schnell included a chimney swift flying in still air at fifteen to twenty-one miles an hour; some kingbirds at twenty-one miles an hour; and cedar waxwings at individual speeds of twenty-one, twenty-three, and twenty-nine miles an hour with little or no wind blowing. Marsh-dwelling red-winged blackbirds flew at seventeen and twenty-three miles an hour; a spotted sandpiper at twenty-five miles an hour. English, or house, sparrows were among the slowest fliers at sixteen to about nineteen miles an hour.

If the English sparrow is one of the slow fliers among birds, which is the fastest of all? I have only one source for the swiftest straightaway flight ever reported for any bird.

Beginning in 1922, E. C. Stuart-Baker, a British ornithologist, lived in the Cachar Hills of India. He was there during a survey prior to the building of the Assam-Bengal Railway. With several companions he made repeated measurements of the flight speeds of the spine-tailed swifts that live in that part of the world. He and his companions used stop watches to check the time elapsed for the birds to fly over an exactly measured two-mile course.

Each day the swifts flew directly over Stuart-Baker's bungalow at Haflang. During these flights they traveled in a straight line to a ridge of hills two miles away then dipped out of sight. The speediest among them covered the distance in the remarkable time of 41.8 and 32.8 seconds, or at the rate of 172 and 218 miles an hour.

The accuracy of this record has been doubted by some British authorities, but until it has been either verified or disproved, it is interesting in suggesting the potential speed at which the fastest birds might fly.

CHAPTER XI

STRANGE AND WONDERFUL
MANEUVERS IN THE AIR

For half an hour we had been climbing the steep mountain trail in upstate New York that led to the wild falcon's eyrie. Bert, my companion, stopped to catch his breath. I was glad that he did. I was breathing heavily from crawling over huge boulders and grasping the trunks of hemlock, birch, and wild cherry to help pull my weight upward.

We had stopped at an opening in the trees through which we could look straight across six hundred feet of gap to the sheer cliff wall opposite. We were standing at the edge of a gorge worn deep into the side of the mountain by a stream that had begun its work thousands of years ago.

Now and then the March wind filled the canyon with its muffled roar. Above it, the faint thunder of waters pouring through the bottom of the gorge reached us from far below. There the stream boiled out of the canyon into the placid river that wound along the base of the cliff. Far up the valley, the river disappeared around a bend.

A puffing steam engine, pulling a line of freight cars tiny as toys, crawled along the railroad tracks on the opposite bank one thousand feet below. We felt a sinking in our stomachs looking down at that height and a powerful urge to leap out into space and float like winged eagles or falcons, high over the red barns and brown fields of the long valley below.

Suddenly we heard a wild wailing. It began low-keyed, then rose higher and higher until it filled the canyon with sound. Bert turned, his eyes gleaming. "The peregrines!" he said softly. I nodded. My heart beat faster. Often The Princess uttered that thrilling cry when she was hungry, or when a wild hawk hung high above her perch, a speck in the blue sky.

We dropped to our bellies and crawled through the dense young hemlocks to the edge of the gorge. The eyrie was not far above, and the wailing had come from close by. We knew what that particular sound meant. It was falcon "talk." We had learned the meaning of it after weeks and months of studying wild peregrines at their eyries from the Appalachian Mountains in Tennessee north to the Canadian border.

Yesterday we had climbed the trail at seven o'clock, before the peregrines had had their first meal of the day. Hidden among the trees we had watched the two mated birds sitting quietly in the early morning sunshine, perched not far apart on a dead hemlock that jutted from the top of the cliff just above the nesting ledge.

Most peregrines mate for life, or until one or the other of the pair is killed. Then the survivor may seek a new mate and not always with success. We remembered one lonely male that had kept his vigil at a Pennsylvania cliff all one summer. That spring he had been unable to get a female to join him, though we saw him try to coax at least two of them to his nesting ledge as they flew by in their migration northward. But all his wailing and flying excitedly from one ledge to another had no effect. Long after the females had gone, he still fluttered his wings on his nesting ledge and called, looking with lingering sadness to the north where they had disappeared. The following year a pair of mated peregrines occupied the cliff. After studying them closely, we were sure that the male of the pair was the lonely peregrine of the previous year that had finally found a mate.

Peregrines are especially devoted during their courtship. In its early stages, the smaller male takes the lead in which he begins the nesting cycle by coaxing the female into scraping a shallow depression in the earth of the nesting ledge. When she lays the almost round three, four, and sometimes five chocolate-spotted eggs in the dusty hollow, they are safe from rolling off the cliff.

For twenty-eight days the peregrines take turns in the incubation until the eggs hatch. They also share in hunting for food, and in feeding and caring for the almost helpless downy white chicks. When the young peregrines are six weeks old and fully feathered, they leave the nesting ledge for the first time. Then they begin to practice flying at living prey.

We were at the cliff in mid-March, before the eggs had been laid. The male was now the aggressive leader in the courtship and he seemed eager to please his mate. At this time, he often hunts alone and brings food back to the female, but sometimes she will join him in a chase that is thrilling to watch.

Yesterday we had seen the hungry pair drop out of the hemlock and sweep down the canyon in a blazing dive. A blue jay had started to fly across the gap. Too late to return to the shelter of the trees, it saw the downrushing peregrines and turned aside.

The male, in the lead, must have known that his mate was close behind. As he often does, he allowed her to make the first strike. At the last moment, instead of taking the bird, he rocketed upward above it. The female's stoop had been aimed well in back and below the jay. As she swept under the bird, it turned away, but she altered her course and followed it like a plane-destroying missile. As she reached the jay, she turned on one side, stuck up a big taloned foot, and snatched the bird out of the air. Then both peregrines flew back to the cliff. Had she missed the jay, the male, poised straight overhead, would have turned over and downward to strike the bird out of the air. This was cooperative hunting, deadly sure, and at its thrilling best.

Today the wailing rose higher and higher in pitch, and then we saw the male flying up the canyon toward us. He was carrying a flicker—a large woodpecker—in his talons. He held the bird well back and trailing, not close up under his breast as many hawks clutch their prey in flight.

The female screamed and the male answered. Now she flew eagerly out of the tree and downward, as though to strike him. As she passed under the male, she turned on her side and snatched the flicker from his talons.

We were to see the young falcons on their first flights at this eyrie perform this same trick of sweeping by under the adults and snatching food skillfully away from them. The marvelous adroitness of these birds in the air was almost unbelievable. Even the small rough-winged swallows, darting and twisting in the gorge below, were not safe from the plunges of the peregrines which sometimes took the swallows out of the air while in full flight.

The female flew to her perch on the dead hemlock. As she plucked the flicker before eating it, its spotted breast feathers and yellow-lined wing feathers drifted away from her on the wind. While we watched her feed, we realized that her mate had disappeared. Perhaps he had gone off into the valley to hunt.

When the male returned in about twenty minutes, he carried no prey. He flew rapidly up the canyon, then climbed steeply to alight at the nesting ledge. He wailed briefly, and sat quietly for a few minutes. Then he began walking about, his head lowered, his shoulders hunched, uttering a low, seductive *wi-chew! wi-chew!* The call was quite different from the wild wailing. It had a coaxing quality. However, the female did not fly to the ledge but sat impassively on the dead hemlock, preening her feathers.

Suddenly the male left the cliff and began soaring and swooping about over the gorge. Bert clutched my arm fiercely. "The courtship flight!" he whispered. Then began one of the most remarkable exhibitions of flying I have ever seen.

The male began his flight in a slow figure-eight, back and forth in front of the cliff and below the female. Suddenly he turned and climbed with rapid beats of his wings. He turned over on his side, came down in a long loop, and began his figure-eight vertically, moving up and down in long curves that carried him high above the cliff. At the top of one of these loops, he turned away, riding the wind down the canyon like a falling arrow. Out over the valley he shot upward and, flapping his wings rapidly, climbed straight up into the blue sky. Higher and higher he rose, turning and twisting and rolling over and over in the air. At the height of his climb, when he was only a black speck, he made a sudden turn that brought him facing downward in a long slanting dive toward the cliff. As he came closer and closer, we began to hear the wind roaring through his wings with a sound like ripping canvas.

Up the canyon he sped, into the wind, and moving so fast that he was a black streak against the gray wall of the cliff. Now he began to shoot up and down, "saw-toothing" like a bucking bronco, up in a long climb, down steeply, then rolling over and over like a whirling leaf. Suddenly he shot skyward, turned with the wind, and let the gale carry him far out over the valley again. Now a tiny speck high in the sky, back he came. He flexed his wings rapidly and fell toward the cliff like an avenging fighter plane swooping on an enemy. His wings were drawn back along his sides as he entered the gorge and leveled off with that ripping sound of the wind rushing through his pinions Suddenly he bounded straight up into the sky and turned over on his back in the beginning of a loop. Down he dropped and straightened away, then bounded upward and rolled forward in another loop, then another and another in quick succession.

As suddenly as it began, his courtship flight was over. He soared in slow narrow circles just below the nesting cliff, then, with another burst of wing flapping, climbed straight up and alighted on the nest ledge. The female, perched in the dead hemlock, was watching her mate closely. She wailed and flew down to the ledge.

Bert was breathless. He turned to me, his eyes shining. "God!" he said reverently. "Did you ever . . . ?"

"No, never," I said, and I never again saw the courtship flight done so magnificently.

Although the courtship flight of the peregrine is probably one of the most spectacular of all birds, the marsh hawk has an exciting one. I saw it one day for the first time over a Pennsylvania cattail marsh.

I had seen in the distance a gray male marsh hawk flying about over some trees. I suspected a nesting pair in the marsh beyond. When I stole through the woods and came to the edge of the opening, I heard the shrill cries of the male. Through the newly leafed April woods I could see him gliding straight up into the air, then diving earthward in a series of undulations, as though he were riding an aerial roller coaster.

He rose to about one hundred feet above the marsh, almost stopped in mid-air, then, like a spent arrow, down he plunged. At the bottom of his dive, he flattened away in a long curve like the back of the letter C, then shot upward again, and hesitated, then dived earthward again. He was gliding up and down in a series of giant U's (UUŬUUUUUUUUUUUU). At the top of a rise, as he hung in the air, he sometimes twisted over on his side, or did a front somersault before diving down toward the marsh again. It was all beautifully graceful, like poetry written on the air.

Somewhere on the ground, among the cattails or grasses, the big brownish female had probably built her nest. The male was showing off before her, but it is suspected by ornithologists that his courtship flight, in which she sometimes joins him, renews and strengthens the bond between them that lasts throughout the nesting season.

Some of the big, soft-feathered owls have remarkable and noisy courtship flights—surprising in these usually silent-flying birds of the night. Ordinarily, owls hunt without making wing noises. As they flap or glide silently through woodlands or over fields, they are quick to swerve or dive down upon an unwary mouse, rabbit, or other animal of the night. Their wings are broad and long, with slotted primaries (a high-lift wing) like the wings of the high-soaring eagles and hawks that hunt by day. But the wings of owls have a marvelous adaptation that enables them to fly without a sound.

The leading edges of their wings are finely toothed—like the tiny jagged teeth of a small saw. August Raspet, a scientist who specialized in studies of the biophysics of bird flight, believed that these toothed edges on the first flight feather silence any noise of the vortex of air rushing over the flying owl's wings. But during courtship, at least one of the larger owls deliberately makes loud noises with its wings.

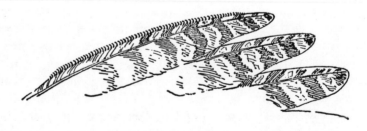

12. *The saw-tooth primary feathers of an owl, which make silent flight possible.*

The short-eared owl of the marshes and prairies, while soaring about at dusk in its courtship flight, glides in easy curves high above the earth. Below him, his mate has built her nest on the ground in a slight depression, lined with weeds and grasses. While soaring, the male gives his courtship "song"—a series of *toots* repeated fifteen to twenty times in an even, low-pitched voice. Suddenly he plunges earthward in a shallow dive, accompanied by a peculiar fluttering noise.

One ornithologist, mystified by this sound, finally discovered how it was made. He watched a male short-eared owl closely. As the owl began his short dive of the courtship flight, he brought his wings beneath him, stretching them backward and striking them rapidly together in short strokes. He ended this wing clapping at the bottom of his dive. The ornithologist said that it seemed as though the owl were applauding his own performance.

Some birds make "wing music" during their courtship flights and one even produces courtship sounds with his tail feathers. I shall never forget one April evening more than thirty years ago when I stood at the edge of a balsam and tamarack swamp in northern New York State, *listening* to a courtship flight I had never heard before. I had long ago heard the musical whistling of the wings of a woodcock in his courtship flight. It came with his rapid descent from an upward spiral into the night sky. And I knew the distant spring thunder of a ruffed grouse's wings as he started his courtship drumming with a deep-toned *thump*, his wings striking the air as he stood on his drumming log, sending out a sound like the muffled beating of a great heart. But this sound was different.

It came from high in the air—a throbbing, bleating *whoo-whoo-whoo-whoo-whoo* that echoed weirdly over the swamp. It was a soft, yet penetrating sound that seemed to come from all parts of the sky.

I searched the air and finally saw in the dusk a small bird with pointed wings and a long bill, flying at a great height. It was traveling rapidly in an immense circle high over the darkening bog. It was a common snipe and the sound was not uttered by the bird. It came from air rushing through the snipe's stiff outer tail feathers which it spread as it flew. Scientists believe that the snipe does this deliberately to warn other males away from its established territory.

101

One day in Shu Swamp on Long Island, New York, I heard a similar sound that came from the movement of a bird's wings in flight. On Easter Sunday, April 6, 1958, I saw two snow-white mute swans (a big waterfowl introduced into America from Europe in New York State in the mid-1800s) flying through the swamp woods. The large male, his long neck outstretched, was leading the female as the mated pair followed the twisting course of a woodland stream. This was not a courtship flight, but as they pumped their seven-foot spread of wings in leisurely strokes, I was deeply impressed with the music of their flight. I found I could exactly imitate the sound of their stiff primary feathers beating the air by whispering "W-a-h! w-a-h! w-a-h!" steadily in the same rhythm as their wing strokes.

The courtship of the nighthawk, a dark, long-winged insect-eating bird that flies about in the spring and summer dusk, is an aerial performance. The nighthawk (not a hawk at all) is closely related to the whippoorwill. The female nighthawk lays her eggs on the ground of an open pasture and on the low flat roofs of buildings in towns and cities, especially in the South.

The male rises to a considerable height over the nesting place, then falls swiftly toward the earth. He plunges head downward, his wings partly closed at his sides. When he seems about to crash into the spot where the female is brooding her eggs or young, he moves his wings down below his body and shoots upward. As he does, the air rushing through his primaries sounds a loud, hollow, *wh-o-o-o-m!* I can imitate this sound by blowing across the bunghole of an empty barrel, or the mouth of an empty jar, bottle, or jug.

The most charming and most delicate of all courtship flights are those of the larks, bobolinks, indigo buntings, and certain other songbirds that often perform over sunny fields. Rising high in the spring or summer air, they do not flap but float about on shimmering, trembling wings while uttering a burst of wild sweet notes that cascade toward the earth.

One of the most spectacular, and possibly most mysterious kinds of bird flight I see practiced every autumn along the coastal beaches from New England south to the Carolinas. As I walk along the strand, I disturb flocks of semipalmated sandpipers and other shorebirds feeding on the wet sandy beaches ahead. Flying up, they dart away over the ocean in a tightly knit flock, turning and twisting in amazingly swift movements in which thirty, forty, fifty or a hundred birds move as one.

Now their white breasts flash over the gray waters, then the flock almost disappears as they turn their dark backs toward me and are lost against the waves. Like the turning face of a mirror they move close-packed, now light, now dark, until they finally turn shoreward and alight like whirling leaves on the beach ahead.

Many observers believe that the leading bird in these flocks—the one flying out in front—is the leader and that the others follow his every motion and change of direction. But some scientists have discovered that the leading bird among sandpipers, pigeons, or of other birds that often fly only in groups of their own kind, does not always lead. They have observed that the leadership sometimes seems to sweep through the flock from along one edge of it, rather than from the front.

I have heard some people say that they believe that the birds, as they fly, are guided by telepathy (thought transference) or that possibly a "common soul" dominates the flock in its perfectly coordinated flight. Whatever guides the flock, the twisting and turning is so swift, so perfectly timed, that a mythical explanation seems the only one to some people. Perhaps scientists will someday find the answer.

CHAPTER XII

DANGERS OF BIRD FLIGHT

Watching birds fly, one gets the impression that they are complete masters of the air. Usually they are—the swallows in their swift, daring flights, skimming low over fields and ponds; the golden eagle in its deadly aerial plunges after its prey; the ruffed grouse thundering through the forest, dodging in and out of tree trunks and branches to avoid the hunter's shot. To suggest that birds have accidents caused by their own misjudgments is difficult to believe after watching their wonderfully skillful actions and split-second timing in flight.

But birds do have accidents. Some are caused by their own imperfections; others by man-made hazards with which a bird cannot cope. Some come with the sudden and tremendously destructive forces that nature unleashes at times on all creatures of the sky and earth.

White pelicans are one of the largest of North American birds. These fifteen- to seventeen-pound snow-white birds have an eight- to ten-foot wingspread. They are magnificent in flight when flocks of them soar together in circles high in the sky. Their white wings, with black outer parts, have well-formed wing-tip slots that remain open as they fly. Like the broad wings of swans and geese, theirs is a compromise in shape between the long but narrow wings of the oceanic soaring albatrosses, frigate birds, tropic birds, gannets, and gulls, and the ultimate in opening wing-slotting of the broader-winged eagles, vultures, and soaring hawks.

In America, the white pelican nests largely on islands in lakes of the western United States. They seem to love aerial exercise. Often, when black thunderstorms loom on the horizon, the adults are stimulated to spiral up into the windy, darkening sky. Then they dive wildly toward the ground. As they rush earthward on rigid, half-closed wings, the roar of air through their stiff primaries sounds like the boom of distant thunder.

On April 4, 1939, a series of violent electrical storms struck Nelson, Nebraska. Roads were flooded, and a farmer of the area, Emil Schlief, sent his fourteen-year-old son Arthur on horseback to the country school to pick up Arthur's younger brothers.

At about 3:30 P.M., as the boy rode his horse along the road, he was stunned by a vivid flash of lightning and an explosion. An electrical charge had struck in an open field about a hundred yards away. As Arthur looked up into the sky, he saw a large flock of white pelicans flying about a hundred feet above the field. Many of them were tumbling toward the ground. The bolt of lightning had struck down thirty-four out of the flock of about seventy-five birds. One soon revived and flew up and away, but thirty-three lay dead, their white feathers singed brown where the heat of the lightning had burned them.

John James Audubon, pioneering American ornithologist, and our first great painter of birds, saw two nighthawks killed by lightning. During one of his Florida wilderness trips in the 1800s he wrote:

> While at Indian Key . . . I saw a pair of these birds (nighthawks) killed by lightning while they were on the wing during a tremendous thunderstorm. They fell on the sea and after picking them up, I examined them carefully but failed to discover the least appearance of injury on the feathers or in their internal parts.

One day my friend Alexander Sprunt, Jr., an ornithologist, and for many years superintendent of the southern sanctuaries of the National Audubon Society, got a well-authenticated story about birds and lightning. It came from the South Carolina Low Country, not far from Charleston.

Cormorants are dark, fish-eating birds of our Atlantic and Pacific coastal waters. They are about three feet long, have a four- to five-foot wingspread, and weigh four to five pounds. Their flight is heavy and heronlike. In the slow flapping of their broad wings, and in their long slender outstretched necks and occasional V-shaped flight formations, they resemble flocks of black geese.

On April 11, 1941, four farmers were talking in a crop field on

Wadmalaw Island, when a violent electrical storm with hail struck at 2:00 P.M. A flock of birds was flying over and at at that moment a sharp flash of lightning came, followed by a terrific clap of thunder. Four out of the flock plunged toward the earth and struck the ground. The men picked them up and found that all were dead. They were double-crested cormorants in spring plumage, and their new feathers were unmarked by the lightning stroke.

Apparently records of people seeing birds killed by lightning are rare, although it may happen oftener than the few reports show. I have one other description of birds killed in flight by lightning.

Dr. John T. Zimmer, whom I knew when he was curator of birds at The American Museum of Natural History in New York City, told the story. He was watching a large flock of migrating snow geese and blue geese in the Midwest when a sharp flash of lightning streaked through the birds in the sky. Many of the geese started to fall, but some recovered before striking the ground and continued on their way. However, more than fifty were killed, one was very badly mangled, and several others had severe internal injuries from their impact on striking the ground. None showed any scorched feathers.

Other birds have met violent deaths in the air from natural forces in a different way. In April 1951, a group of golden-crowned kinglets, winter wrens, a sapsucker, robin, and a purple finch were part of a large number of songbirds migrating northward over Mount Washington in New Hampshire. Suddenly a powerful downdraft of wind carried them below the ridge and hurled them violently against the mountain slope. The next day an ornithologist found their broken and battered bodies lying on the snow.

Tropical hurricanes that move northward, then westward across the eastern United States, often carry seabirds far inland from the ocean. Many oceanic birds can take off only from water on which they get a running start with the help of uplifting air currents. Some are borne aloft by leaping into the air from a cliff above the water. When oceanic shearwaters and petrels, carried inland on hurricanes, alight, they cannot take wing from the ground. Usually they starve or are killed by dogs, cats, and other predators.

One night in November 1954, many migrating pied-billed grebes —waterbirds that look somewhat like small ducks, and are often called "hell-divers"—were flying over Lumberton, North Carolina, when they got into a similar predicament. Grebes, too, can take off only from water by running rapidly over the surface and vigorously flapping their short wings. When they alight on land they are helpless. During a rainy night, the migrating grebes mistook a wet, shining North Carolina highway for a stream. They alighted on the road and were unable to take off. The next day, heavy automobile traffic on the highway killed large numbers of the helpless birds.

One of the most remarkable "natural" flight accidents I have ever heard of was that of a pheasant in England that was killed by flying into a flock of starlings. Hawks, which sometimes prey on flocks of starlings, are often thwarted when the starlings, flying in a loose flock, suddenly fly close together to present an almost solid mass of birds to their attacker. Usually the hawk will turn away from the flock at this maneuver and fly off to try to find a single bird.

Apparently many hawks have learned that it is dangerous to fly into a flock, or perhaps their instincts warn them that they cannot maneuver among the closely flying birds. The pheasant, frightened into quick flight by a hunter, was so intent on escaping that it flew directly into the starlings.

Not long ago, I learned of another remarkably strange accident to a bird. This one occurred near Lawrence, Kansas. Small birds will sometimes "gang up" on a larger bird, such as a hawk or an owl, that they consider an enemy. They dive down on the larger bird's back, plucking out a few feathers or striking with the bill and feet at the bird's head. Ornithologists call this behavior "mobbing." Its aim and usual effect is to harass the larger bird until they drive it away.

Two ornithologists from Lawrence one autumn day saw an osprey, or fish hawk, flying over an oxbow lake of the Kaw River. It was flying wildly and erratically, pursued by smaller birds that were diving at it. Frantic to escape its small, mobbing pursuers, the osprey dodged, and turned over several times on its back. Each time it struck upward with its talons at its assailants as any attacked hawk will do. The action reminds one of a cat that will turn over on its back and strike up at its attacker with its claws.

Suddenly the osprey tumbled downward and fell into the lake. The two men, after considerable delay, finally reached the bird. It was floating on the water, still alive, but it died a few minutes later of shock.

On examining the osprey, the men found that during its aerial twisting and clawing at its tormentors, it had accidentally sunken one of its razor-sharp talons deep into one of its wings. The impact had broken its arm (humerus bone) above the elbow. The big hawk's violent efforts to pull its talon from its wing had been unavailing. It died, helpless and unable to fly, a victim of a trap of its own making.

Single birds sometimes collide with each other in the air. One day on Long Island, New York, I saw two tree swallows in a flock come together while swooping over a pond. Only at the last moment did they escape certain injury or death by veering sharply away from each other.

On June 8, 1949, Dorothy Snyder, a distinguished ornithologist from Massachusetts, saw two rough-winged swallows collide in the air over a pond in Berkshire County. One of the swallows fell to the water and made no effort to rise. When Miss Snyder retrieved it the bird was dead.

Many natural accidents are caused by fright or by mistakes in a bird's judgment. Dr. Arthur A. Allen, Professor of Ornithology at Cornell University, made pioneering studies of diseases in ruffed grouse at Ithaca, New York, in the 1920s. Later he told me that he had autopsied dead grouse that had had violent accidents during their lives. One had a fair-sized twig within its body, evidently jammed down the bird's throat as it flew at high speed through the undergrowth. Another had part of its crop containing acorns torn away and pushed under the skin of the lower breast. It had obviously struck a tree or other hard object with terrific force. The crop itself had healed perfectly.

On November 24, 1939, David Nichols of Berkeley, California, saw a varied thrush that refused to fly. When he picked up the helpless bird, he discovered that it had pierced an acorn with its bill. Unable to free itself from the acorn, the thrush had apparently been unable to feed itself for a long time. Nichols pried the acorn loose and the bird flew weakly away.

Apparently all the natural accidents to birds, when combined, cannot equal those caused by man-made hazards. Tens of thousands of flying birds, darting across highways, are killed each year by speeding automobiles; many more are killed or crippled in migration at night when they fly into 800- to 1000-foot-tall TV towers and their cables. There is an often appalling destruction of migrating birds, forced to fly low at night by clouds or rain, that strike against tall city buildings. Thousands of others are blinded and "crash land" because of the dazzling, 25-million-candlepower lights of ceilometers at airports. These are beamed upward periodically to measure cloud heights.

One of the most innocent-appearing but deadliest hazards to flying birds are the thousands of picture windows and other large glassed areas of suburban and country homes. Birds feeding and nesting in the garden see trees and shrubs reflected in the window glass. Apparently they fly into them under the impression that they are flying into an extension of the garden itself.

Some homeowners try to prevent this annual killing of bobwhite quail, catbirds, cardinals, sparrows, mourning doves, and other of their garden birds. They install almost invisible nylon marquisette netting across the window glass. This cushions the force of impact of any bird that flies against the window and saves many bird lives. Other homeowners have attractive awnings over the windows and keep them lowered. This discourages birds from flying against the glass.

Barbed-wire fences are another deadly man-made hazard to birds. Not long ago, I heard of a king rail that was caught on a strand, and a western great horned owl that died with its wings hopelessly entangled in the spurs of barbed wire.

One day, while I was editor of *Audubon Magazine,* a woman brought to my office in New York City a crippled short-eared owl. The bird had been impaled by one wing on a TV aerial of a housetop in Jamaica, New York, as it flew over the house toward the nearest marsh. Its wing was so badly injured that it never could fly again. It lived for ten years after that as a household pet.

But of all man-made threats to birds, utility wires can be the most fearsome. A friend of mine found a shorebird in one of our western states that had both of its wings cut completely off when it flew into a wire. A golden eagle in Minnesota, chasing a bird, was so fiercely intent on its prey that it flew into utility wires and was electrocuted.

One day my peregrine falcon, The Princess, gave me a terrible fright. I was flying her in an open field and she suddenly turned and flew swiftly toward some wires strung between poles along the edge of a road. Apparently she did not see them, but she flew between several rows of them and came out uninjured, except for a few broken feathers. It was a narrow escape.

On May 24, 1965, a California condor, one of the rarest birds in America, was killed in Fresno County, California, when it collided with a high-tension power line during its soaring flight. In November 1965, a whooping crane, an even rarer bird at that time, was killed when it flew into a high-tension wire near Ludell, Kansas. The crane was one of a flock migrating from its nesting grounds in the Canadian Northwest, southward to its wintering grounds on the east coast of Texas.

Fortunately, not all birds strike utility wires. Many of them see the wires in time to escape striking them; others flying about in their home territories every day, learn by experience that the wires are there and avoid them. A remarkable story about birds and utility wires came to me from a friend in England. It illustrates not only the keen eyesight of birds, but their marvelous maneuverability and split-second timing in the air.

A European sparrow hawk, an accipiter, or bird-killing hawk related to our American Cooper's hawk, and looking much like it, was flying full speed after a European finch. The finch, squealing in fright, flew directly toward a mass of utility wires that stretched across an open space between two poles. At the last moment, just before striking the wires, the finch dropped like a stone into the safety of a garden shrub on the ground below.

The sparrow hawk, intent on the finch, did not see the wires until it was about to hurtle into them. To save itself from being cut to pieces, it shot straight up into the air. When above the wires, it turned over on its back. For a few seconds it flew upside down, back in the direction from which it had come. Then it righted itself and flew rapidly away.

My English friend said that it was one of the swiftest recoveries, and perhaps the most brilliant, he had ever seen made by a bird in the air. The hawk's split-second change of direction had saved its life.

CHAPTER XIII

FLYING UNDER WATER

During more than forty years of his long life, Arthur Cleveland
Bent, an American businessman turned ornithologist, wrote twenty-
one volumes under the title *Life Histories of North American Birds*.
When he traveled along the coast of the Pacific Northwest early
in this century, he was awed by the tremendous numbers of seabirds of the auk family that he saw there. Birds of this family live
only in the Northern Hemisphere and are especially numerous along
the Alaskan and Siberian coasts. Mr. Bent was greatly impressed
that they not only flew through the air as other birds do, with rapid
beats of their small wings, but flew into and under the water in
their search for food with scarcely a change in pace.

Beginning at Three-Arch Rocks, Oregon, where about 750,000
common murres, a sixteen- to seventeen-inch-long member of the auk
family, nest on the rocks each year, Mr. Bent moved northward on
a ship traveling along the Alaskan coast. It was the middle of June
when he saw the greatest concentration of nesting birds in all of
North America. For hundreds of miles along the rockbound shores
he was seldom out of sight of birds that blackened and whitened
the cliffs and pinnacles of the islands along the shore.

Of the auk family,* there were millions of the black-backed, white-
bellied common murres, probably the most abundant of all seabirds
of the North Pacific coast, and thousands and thousands of the
smaller pigeon guillemots, tufted puffins, horned puffins, several kinds

---

* In North America, the auk family includes murres, the guillemots, murrelets,
the extinct great auk, auklets, puffins, dovekies, and the razorbill of the North
Atlantic coast.

of murrelets, and four or five species of auklets, of which the six-inch-long least auklet, is the smallest in the auk family.

The black and white murres (pronounced *murs*) are the gentlest of all the auks. These sharp-billed seabirds live in closely packed colonies of their own, but tolerate related species, and live with them among the rocks in complete harmony.

Mr. Bent noted that the murres flew through the air swiftly and strongly with rapid beats of their small wings. They were so heavy-bodied in proportion to the size of their wings that they could not rise off the water without running rapidly along the surface. Swans, geese, loons, grebes, some of the ducks, and other heavy-bodied birds also cannot fly into the air from the surface of the water without a running start.

In flying from their nesting cliffs, the murres glided swiftly downward at a steep angle, then swept outward in a long curve before leveling off. After diving under the surface, they flew rapidly through the water and with repeated swift strokes of their wings, herded small fishes into a group. Then they darted among them, snatching one here and there in their sharp-pointed bills.

Mr. Bent noted that the eight-inch-long Cassin's auklets, which nest from the Aleutian Islands of Alaska, south to Baja California, dived under the sea and could stay under two minutes or more. They moved swiftly under the surface by pumping both wings, just as the least auklet and other auklets do. Mr. Bent watched them fly through the water with their wings held parallel with the body, and not outstretched as in their aerial flight. And like the common murres, they moved their wings straight out from their sides with each stroke, then back toward their sides again in a flaplike motion.

The fourteen-inch-long pigeon guillemots (GIL'-eh-mots) flew strongly and directly through the air, but usually close to the water. When they dived under the sea, they, too, used their wings to propel them, holding their bright red feet trailing out behind, probably to help them in steering.

But of all the birds of the North Pacific coast that fly under water, Mr. Bent was most charmed by the tufted puffins. These blackish seabirds, about the size of the pigeon guillemot, have big, bright red, parrotlike bills (males and females are marked alike) and a comically solemn appearance from which they are called "sea parrots." Each has long straw-colored plumes that flow back from above the eyes, and white face, giving it another of its common names—"the old man of the sea." They are not shy, and on the Arctic islands where they nest, they spend much time sitting upright at the entrances to their underground nesting burrows, or at crevices in rocks from which they peer out curiously.

I wanted to see exactly how these comical birds swim under the surface of the water, and so, one day, I visited the Aquatic Birdhouse in the New York Zoological Gardens (Bronx Zoo) in New York City. The zoo officials had several pairs of tufted puffins confined in an indoor glass tank where one could see them swimming and diving under water.

As the puffins swam about on the surface, they sometimes rose up and fluffed their feathers and beat their short wings like ducks. They have webbed feet, and as they swam about they paddled like ducks, thrusting first one foot backward and then the other, spreading the webs between the toes with each alternate stroke of a foot.

When the puffins submerged to swim under water, they nosed forward in a little dive and propelled themselves with strokes of both wings together. This sent them ahead in a lunge with each wing stroke, down, then up, down, then up, with their feet dangling in back but not paddling with them. (Mr. Bent, from the deck of his ship sailing along the Alaskan coast, noted that tufted puffins, when swimming below the surface, not only beat their wings in unison but paddled with their feet.)

When trying to fly from the surface, tufted puffins, with their relatively heavy bodies and small wings, have difficulty getting into the air. Mr. Bent often saw them make futile attempts to fly from the water, flapping furiously along the surface only to drop back and try again. They have equally great difficulty taking off from land. They usually get into the air by launching themselves out over the water from a cliff or hillside. Once under way, however, their flight is strong and sustained. They eat fishes up to ten inches long, also mollusks, sea urchins, and algae. Their feeding waters are often many miles from their nesting colonies, thus making it necessary for them to take one long round trip each day.

Before I left the Bronx Zoo that day, I stopped at the Penguin House to see birds that can "fly" under water better than any others. Penguins are not related to the auks, but are a family of flightless seabirds that live only in the Southern Hemisphere. Dr. William Beebe, the American scientist and writer of many delightful books about birds and other animals, considered penguins to be the most wonderful birds in the world. There are seventeen or eighteen species. Two live in Antarctica, and many others around the southern coasts of South America, and north along the west coast to Peru. Others live along the southern coasts of Australia and Africa. Although penguins cannot fly through the air, scientists believe they are descended from birds that could. Apparently they have lost their ability for aerial flight in the adaptation, through millions of years, to life in the sea.

Inside the cool, dark building, lights brightened a pool of greenish water beyond the glass. On the far side of the pool, several big king penguins, about three feet tall, and each weighing about seventy pounds, stood on a rocklike surface with their flippers held out from their sides. Instead of having a certain number of feathers on each flipper, divided into a well-marked series as on the wings of other birds, the paddles of penguins are covered thickly with small feather scales. The rigidity of the flipperlike wings, along with their rotary motion, which I was soon to see, makes their resemblance to the propeller of a ship an apt one. With these propellers, penguins fly through the water with almost the identical wing motion of a bird in the air.

I noticed that the bodies of the king penguins were covered densely with sleek, scalelike feathers and that their bodies were torpedo-shaped. Scientists have estimated the speed of some penguins swimming under water at twenty-five miles an hour.

Two other kinds of penguins, much smaller than the king penguins, were swimming about in the pool. One was the Humboldt's, named for the German scientist Alexander von Humboldt, the great explorer of South America; the other, the rockhopper penguin.

The handsome Humboldt's, or Peruvian, penguin, common along the coasts of Peru and Chile, is about two feet long and weighs about nine to eleven pounds. I saw that both sexes looked identical as they swam along the surface of the water within a few feet of

13. The under-water flight of the Humboldt's or Peruvian penguin.

where I stood. They had black heads with a line of white feathers
well above each eye. The line extended down and around the black
throat of each bird, and ran down their sides as a white stripe that
contrasted with their black-feathered backs. Their breast feathers
were snow white, the irises of their eyes, reddish brown.

I watched them swimming about on the surface of the pool,
paddling with their webbed, well-clawed feet. When they went under
the surface, they did not lunge forward in a dive, but simply put
their bills and heads below the water and then sank down. At the
same time they began to paddle under water with their wings, holding
their webbed feet trailing out in back under the sharp-pointed tail. As
they moved under water, they flapped their wings, or "flippers" si-
multaneously, and rowed with a graceful, rhythmic forward and backward motion of the wings. With each stroke they seemed to rotate
their wings from the shoulders.

When swimming about under the surface of the coastal waters of
Peru and Chile, the Humboldt's penguin darts through the water,
making lightning-swift turns by braking one wing, while beating the

other. Looking down from the deck of a ship, one can see them snatching and devouring fishes. Sometimes they remain under water for fifty to seventy seconds, and after coming up for air, dart below the surface again in their search for food. They are especially fond of anchovies.

The under-water fishing of penguins is not always safe for them. Many are killed and eaten by a deadly enemy, a seal of the Southern Hemisphere called the "sea leopard." Males may be nine feet long and females up to fifteen feet and they have voracious appetites. From the stomach of one of these seals killed at Bay Isles, South Georgia Island, Dr. Robert Cushman Murphy of The American Museum of Natural History, New York City, took the remains of four big king penguins.

Perhaps any bird that alights on the water or dives below the surface may be attacked by enemies that live there. Some years ago, I heard a story told by several duck hunters who were crouched one fall morning in a rowboat on Raritan Bay, New Jersey. They were about five miles off shore. Suddenly, in the dim morning light, they saw a large creature threshing wildly about on the surface of the bay. Drifting closer, they recognized it as an angler fish that appeared by its struggles to be in mortal agony. A bunch of feathers protruded from its mouth and it seemed unable to sink.

Grasping the three-foot fish, they hoisted it into the boat and found a large herring gull stuck in its throat. Despite its big mouth, about ten inches wide, the fish had been unable either to swallow or to disgorge the fair-sized gull. When forced from the fish's mouth, the bird was dead but it still had its head tucked in back of one wing. The angler fish had risen to the surface and had caught the gull as it floated on the water asleep.

On November 19, 1965, a fisherman of Vineyard Haven, Massachusetts, reported to the Bird-banding Office of the United States Fish and Wildlife Service in Washington, D.C., that he had recovered an aluminum band from the leg of a herring gull. The gull had been swallowed by an angler fish. About two years before, in the summer of 1963, when the gull was a flightless baby in a gull colony near Edgartown, Massachusetts, a woman birdbander had put the identifying ring around the bird's leg.

The angler fish, which in American waters frequents the Atlantic Ocean from Newfoundland to Cape Hatteras, is big and bulky and looks like a large catfish. The inside of its mouth is as ferocious-appearing as a shark's. In both jaws its long, sharp teeth are hinged to depress backward toward the angler's throat. This allows the inward passage of any creature the angler fish tries to swallow. But any outward pull on the teeth brings them forward and erect, thus hooking the living prey in the angler's mouth and preventing it from backing out.

Sometimes called "goosefish" for its attacks on wild geese, the angler fish also captures loons, wild ducks, grebes, and auks, of which the black guillemot and razorbill live along the North Atlantic coast. There is an authentic record of an angler fish that when captured and opened up, had the remains of seven wild ducks in its stomach.

Other fishes besides the angler fish attack birds. Late one afternoon, a fishing party in the Atlantic Ocean watched a loon swimming and diving near their boat. Loons propel themselves with their powerful feet, using their wings below the surface only for balancing and steering. When feeding undisturbed, a loon may stay under water about thirty to forty-five seconds, but if frightened or chased by men in a boat, it may stay under three to five minutes and can swim under water for several hundred yards with each dive.

While the loon floated for a moment between dives, the members of the fishing party were astonished to see a big thresher shark come to the surface. It appeared alongside the loon and struck it a tremendous blow with its tail. Then it swallowed the bird in one gulp and disappeared below the surface.

Cormorants, which catch fishes by diving after them, share with loons and grebes the ability to change their specific gravity and to make themselves less buoyant in order to sink part way below the surface. They do this by expelling air from their bodies and feathers. After diving, they shoot swiftly about, propelling themselves mainly by their feet, but assisted somewhat by their half-opened wings. Fishermen have seen cormorants suddenly whisked below the surface with incredible speed—victims of the eight-foot-long monk fish.

Even the edible codfish will swallow birds that swim under water. Some years ago, a large codfish caught off the coast of Newfoundland had swallowed a black guillemot, or "sea pigeon." The black guillemot is about a foot long and is, with the razorbill and Atlantic puffin, one of the auks that nests along the rocky northeastern coast of America.

Most of the birds that use their wings to swim under water—penguins, shearwaters, diving petrels of the southern oceans, cormorants, and auks—are waterbirds, or seabirds. (The diving ducks, or sea ducks, submerge completely with quick, arching dives from the water's surface, but once under water, usually propel themselves with their feet.)

There is one landbird, however, a small songbird, that also "flies" under water. It is the eight-inch-long dipper, or water ouzel, of the mountain streams of our western states, Alaska, and extreme western Canada. Dippers walk in and out of the waters of swift-running streams, lakes, or ponds on their long legs and unwebbed feet, searching for water insects such as the larvae of caddis flies that they find under stones. They also eat water bugs, water beetles, and even small fishes. They swim about under the surface in deep water by propelling themselves with quick strokes of their stubby wings.

The dipper has extraordinary adaptations for its way of life. Its grayish brown plumage is soft and filmy with a thick undercoat of down; it has a relatively tremendous oil, or preen, gland from which it squeezes oil with its bill to waterproof its feathers; it has a movable flap over the nostrils to keep out water; and a third eyelid (nictitating membrane) which it closes over its eyes when it is flying through the spray of falls and rapids of its mountain stream home.

Scientists are not quite sure from what ancestral songbird stock the dippers are descended. Some believe they developed from some thrushlike bird; others that they are descended from a wrenlike ancestor. They are similar to wrens in their looks and habits. They swim expertly under water and have been seen to fly through twenty feet of water to reach the bottom of a pond.

One observer watched two dippers plunge beneath the surface of a pond while flying. They continued their flight under water for the full length of the pond.

CHAPTER XIV

HOW IT ALL BEGAN

14. Archaeopteryx.

In my walks through woods and swamps, or along the bases of cliffs in the mountains, I may find on the ground a lone handsome wing pinion of an eagle, hawk, owl, raven, wild turkey, or some other large bird that has shed one of its flight feathers in its annual molt. Besides admiring the beauty of these and enjoying their faintly sweet odor, I can study their shapes and their rigidity or softness that may teach me a lot about their use to a bird while it was flying or soaring through the air. But I do not always know what species of bird that wore that single feather. Sometimes it requires some hard sleuthing to discover the bird's identity. For me it has become a kind of detective work.

Once a year, practically all adult birds molt, or drop out their old feathers and grow new ones to replace them. Hawks, owls, woodpeckers, jays, warblers, sparrows, and other landbirds molt their flight feathers gradually, or only a few at a time; therefore they can fly during their molt. It would be disastrous if they could not. They would be unable to fly in pursuit of their food, or to escape from their enemies.

However, many waterbirds—ducks, geese, swans, coots, and others—drop their flight feathers all at once, right after the nesting season. They can do so safely because they live on water. For the few weeks that they are flightless, they still can hunt for food in the water, or escape from an enemy by swimming away, hiding among marsh plants, or by diving below the surface.

No bird, however, even when it is capable of strong flight, is always safe from a predatory bird or other animal. In my walks through fields and woods, I sometimes find a small pile of feathers lying on a country road, or along a woodland trail. Perhaps there is only a single feather, or luckily for my success in identifying it, a piece of the bird's wing, or possibly the remains of a leg or foot. From this I can be sure that it was killed by a predator and plucked and eaten on the spot.

But what bird was it, and what had killed it? It is sometimes days, weeks, or even longer before I can identify a bird that has left behind only a feather or two. However, my task is simple compared with that of some scientists I know. Each day of their lives they engage in some of the greatest detective work in the world of birds. They are ornithologists who have become paleontologists—those who specialize in trying to identify birds of the past from their fossilized remains.

Many years ago, for two weeks, I kept close company with Dr. Alexander Wetmore, an American ornithologist who is a leading authority on fossil birds. Dr. Wetmore was then Secretary of the Smithsonian Institution in Washington, D.C., and I was writing a biographical sketch of him for *Audubon Magazine*. Besides his wonderful skill at identifying living birds in the fields and woods, he is an expert in the recognition of birds from their bones. From petrified bits, he has described eighty species; among them tropical chachalacas, giant barn owls, stilt-legged hawks, tiny parakeets, and limpkins (large, rail-like birds now limited to tropical America) that had lived in the forests of ancient Nebraska; a peculiar Ice Age turkey, with a three-pointed spur on its leg instead of a single spike, and a California condor and a larger giant vulture all of which had once lived in Florida.

From his studies, Dr. Wetmore told me that he was convinced that the evolution of our existing species of American birds began in the Miocene and Pliocene periods of our geological history—approximately 10 to 20 million years ago—when the Alps, Himalayas, and other high mountains of the world were being formed.

His skill in reconstructing a complete bird from a bone fragment is like that of a laboratory criminologist supplying vital information needed to solve a crime. Six years before I interviewed him, a paleontologist from the Carnegie Museum in Pittsburgh had sent him the fossilized skeleton of a bird dug up in Wyoming. For months Dr. Wetmore puzzled over the bird's identity. One day he happened to glance at the bird's skull from a different angle and saw a tiny incomplete abridgment of the nasal bones. This identified it for him at once as one of the most remarkable fossil birds he has ever described—a previously unknown, flightless "running" vulture. The bird, about the size of a wild turkey, had lived along the shores of inland lakes on the North American continent millions of years ago.

One of the most exciting bird detective stories I know was solved in this way. From it, scientists were able to reconstruct with accuracy the oldest bird known to the world and to look at the past life of one of the ancestors of all the birds we know today.

In 1861, at Solnhofen, Bavaria, a quarry worker dug up a fossil feather of an unknown bird and its impression on a counter slab of what is called Solnhofen limestone. That same year, in the same quarry, workmen found the fossilized, incomplete skeleton of an ancient, reptilelike bird. It was about the size of a crow, and its fossilized feathers were similar to the one found by the quarry worker earlier that year.

This incomplete skeleton of the unknown bird eventually came into the possession of Dr. Karl Haberlein, the district medical officer. He sold it, with a fine collection of other fossils, to the British Museum in London. Scientists, studying the fossilized skeleton, concluded, from the slate deposits of the Upper Jurassic period in which the reptilelike bird had been found, that it had lived about 140 million years ago. They named it, *Archaeopteryx* (pronounced "are-kee-OP'-ter-icks") *lithographica*, meaning, "the ancient-winged creature of the stone [slate] for drawing."

What was it like and how had it lived?

Scientists knew enough about the life of the geological period in which *Archaeopteryx* had lived, and from the bird's structure, to believe that it was probably a bird of the forests. Its fossilized wing feathers were arranged in exactly the way of those of the true birds, and in structure, identical with those of present-day birds. The wings each had eight primary (flight) feathers, attached to the hand or wrist part of the wing. It also had secondary feathers that grew out of the forearm, and wing covert feathers just as in a modern bird's wing. However, each wing had claws on the "hand" part of the wing —a reptilelike feature—and it had a long, lizardlike, feathered tail.

Judging from the position of its "big" toe (hallux), which was opposite the other three, and adapted to gripping a perch, it had apparently lived in the forests. However, it had no hollow bones as modern, or present-day birds have, and its flight was apparently limited because it had no powerful flying muscles. Scientists were sure that it did not have strong flying muscles because it lacked a keel on its breastbone. In modern birds, this keel is the base of attachment for the great wing muscles of flying birds. *Archaeopteryx* was basically a glider, but also may have propelled itself by its weak wings to fly from branch to branch of its forest home. *Archaeopteryx* also had teeth, ordinarily a character of the reptiles of that time, but some later fossil birds also had teeth.

Its feathers show that *Archaeopteryx* was a bird, although a very primitive one. Up to the present it is the oldest known bird. The second oldest whose remains have been discovered was estimated to have lived about 20 million years later than *Archaeopteryx*. It was a flamingolike bird described in 1931 from a fossil dug up in France from the rock beds of the Lower Cretaceous period. A scientist named it *Gallornis straelini*.

Zoologists are convinced from studies of the comparative anatomy of birds and reptiles, and from the fossil evidence of ancient animal life, that birds, without a doubt, arose during their evolution, from reptiles. Scientists are also now generally agreed that both birds and dinosaurs arose from primitive, unspecialized thecodont reptiles, —animals with many teeth set in sockets. The thecodont reptiles with hind limbs longer than the front ones and a long tail, were believed to have run about over the ground bipedally—that is, on their hind legs in an upright posture. There are two main theories of how these ancient animals developed into birds, or became animals that fly.

In 1907, Baron Francis Nopcsa, in an article "Ideas on the Origin of Flight," published in the *Proceedings of the Zoological Society of London,* proposed that the development of flight came from ground-running, long-tailed bipedal reptiles, which flapped their forelimbs as they ran rapidly over the ground. Baron Nopcsa reasoned that the scales on the forelimbs of these upright reptiles became longer in time and that their rear edges became frayed and eventually evolved into feathers.

Baron Nopcsa also believed that there were three stages in the evolution of flight: (1) parachute, or passive flight; (2), flight by flapping of the wings, or flight by force; and (3), soaring, or flight by skill. He thought that the ancient and oldest fossil bird, *Archaeopteryx,* was still in the first stage of (2) active flight.

However, Gerhard Heilmann who wrote a book, *The Origin of Birds,* published in 1927, disagreed, as did Othniel Charles Marsh, the great American paleontologist who did some of his most brilliant work in the discovery of fossil birds. They and other authorities believed that the land-dwelling ancestors of birds became tree climbers. This was before there was a great difference between their front and hind limbs. Jumping from branch to branch favored the evolution of lengthening metatarsals of the hind limbs and a backward-directed hallux. This enabled these tree-dwelling pre-bird animals to securely grasp branches. The forelimbs, then used for climbing, kept the claws on their digits. The wings, or forelimbs, remained large and were not reduced by evolution as is commonly true of the ground-running (cursorial) animals, which had adopted the bipedal (upright) method of progression.

Each limb, therefore, became adapted to specialized and different uses—rear limbs for leaping or hopping; forelimbs for climbing through trees. Gerhard Heilmann emphasized this *independence* of the two limbs in contrast with those of the Pterosaurs (flying reptiles in no way related to the evolutionary stem of birds) and the flying mammals, for example, bats. In these animals the forelimbs and hind limbs are, and were, connected by a patagial skin fold.

Heilmann also believed that the evolution of feathers from reptilian scales, probably preceded homoiothermism, or "warmbloodedness"* in birds, the condition of a relatively even body temperature as in most mammals, including man.

A German paleontologist, Hans Böker, suggested that the original birds flapped their forelimbs when jumping from branch to branch of trees. However, Sir Gavin de Beer, British scientist, thought it probable that simple gliding preceded flapping because studies of *Archaeopteryx* showed that it had no carina (keel on the breastbone for muscle attachment); therefore its pectoral (flying) muscles must have been weak.

Relatively few complete, fossilized skeletons of birds have been discovered in the world, although numerous bone fragments have been found. These are usually the larger denser ends of wings and leg bones, or parts of the pectoral girdle—the part of the bony skeleton that supports the wings of a bird. Compared with the fossils of other vertebrate animals, the record for birds is incomplete and fragmentary. The bones of birds are light in weight and fragile. They are easily and quickly destroyed and rapidly decompose. Bird fossils have been found in caves, dried-up lakes, diatomaceous earth strata, bogs, rock quarries, tar pits, and Indian kitchen middens.

Dr. Alexander Wetmore believes, from his studies of fossil birds, that, as a group, or class, birds reached their maximum abundance in the Tertiary period of 2 to 60 million years ago.

Up to about 1960, almost 1700 species of fossil birds had been identified, of which about 900 kinds are extinct. The remaining 800 species are representatives of birds still living. There are an estimated 8650, or possibly 9000, living species of birds in the world. All of these living species as well as others that became extinct, probably existed during the Pleistocene period of the last million years.

* Technically, homoiothermic, meaning "constant temperature."

In 1960, Pierce Brodkorb, a paleontologist of the Florida State Museum at Gainesville, estimated that, from the time of *Archaeopteryx*, 140 million years ago, more than 1½ million species of birds have lived in the world. Of these, fewer than 10,000 remain.

CHAPTER XV

EPILOGUE
FAREWELL TO A FALCON

I had to free The Princess, my peregrine falcon. Within six weeks I would be called into the Armed Forces in World War II, and I did not know when I would get home again. I could not leave her with friends. They would not know how to fly her, and I could not allow her to sit through the war on her perch, awaiting my return.

Those powerful, sharp-pointed wings needed daily exercise in the freedom of the skies. Her keen dark eyes, to remain sharp and bright, needed to scan far horizons from her place high above green fields and woods; to pierce the misty miles of some sparkling ocean beach, where the sandpipers and plovers would whirl away beneath her; to search the dark marshes where the ducks would fly up wildly before her hunting.

And so I took her, one Sunday morning in September, to the cliff where she had been hatched on a rock ledge high above a river valley. The parent falcons that had nested there each year were not in sight. Perhaps they had moved on with the migrating waterfowl and the swallows that followed the valley on their way to the great wintering grounds far to the south.

The Princess and I had been together for six years. It would not be easy to let her go. But for a long time she had been superb in her ability to strike any prey out of the air that she chose. She knew all the flying tricks that a falcon needs to know to survive from day to day. Her wild instincts were all there. Although she had gentled with our companionship, her fierceness and deadliness in the hunt were equal to that of a wild falcon. She would survive, I was sure. All

15. The Princess and another peregrine flying away.

that she needed was forty-eight hours on her own to become a truly wild bird again, no longer dependent on me for food and water, and for my companionship in which I stroked her breast feathers and talked softly to her as she looked sharply about or scanned the blue skies for the sight of some wild hawk flying over. With her first wild kill, she would be independent and self-sustaining.

I had placed a government aluminum birdband securely around one of her legs. The band had an identification number and the inscription, "Notify U. S. Fish and Wildlife Service." If she were killed, and the band number reported to the government, I would know her fate. But in all the years that were to follow, I was to have no report of her.

I delayed freeing The Princess as long as I could. I had fed her a half-crop of tender beef. Now she would not be wildly hungry when I released her. She would not be so eager to return, and this would help break her bond to me. It would take only a little while for her to become used to her freedom.

I had walked with her perched on my fist to the top of the rock cliff. Together we looked down the long valley. Below us lay the golden fields and woodlands. Like a silver thread, the sparkling river wound southward to the place where the fields and the shining river and the towering cliff were lost in the September mists.

I cut The Princess's jesses from her legs with a pair of scissors. They dropped to the ground. Now she was completely free. She gripped my gloved fist with her big yellow talons, as though reluctant to go. A light breeze that seemed forever to blow up the face of the cliff touched her breast. She raised her wings slightly. I gave her a gentle toss into the air and she glided down over the rock ledge and away. Swiftly she moved out over the valley, with that familiar quick beat of her wings followed by a short sail, rising higher and higher in tight circles. Then she swept downward below the face of the cliff and disappeared.

I turned and walked down the trail. I knew what The Princess would do. She would go off on a long adventurous exploring trip, perhaps for an hour. Then she would return and hover above the cliff where I had stood. I did not want to be there when she came. She was free now, and when she returned, she would wait only a little while before going off to hunt.

When The Princess had killed and had fed, she would roam farther away. A strange wild peregrine would come drifting by, traveling southward on the trail of the migrating waterfowl and the swallows that were moving down the valley. The Princess would rise to meet the stranger, and together, yet apart, they would drift away until they were lost in the blue haze of the far horizon.

SOME TECHNICALITIES OF
BIRD FLIGHT

Aerodynamics – Birds fly by the same aerodynamic principles as an airplane. They use much the same type of equipment—wings for support and steering; propellers to drive them through the air; a tail to help steer and to help them brake speed of flight and for control when landing; and wing slots and wing flaps to help them take off from the ground or to alight.

Design of Wing – In order to move the bird efficiently through the air, the wing and bird itself are streamlined. The predominant aerodynamic forces that act on an airplane also act on a bird—they are "lift" and "drag." These forces influence the shape of the wing, the basic lifting surface of either a bird or an airplane. If we were faced with designing a wing that would get lift from the air and could "fly," we might start out as follows—we might test the air flow first in a wind tunnel on a blocky, or square object. In the wind tunnel, when the visible smoke stream hits the square object, the air is seen not to flow smoothly around the surfaces of it, and does not close immediately behind it. Instead the smoke stream breaks up and is *deflected away* from the object so that the air no longer presses against the object's sides with the same force. Moreover, the air stream does not close up again behind the obstruction until the air has moved some distance behind it. Therefore *pressure on the rear surface* is also reduced and there remains a disproportion to pressure on the front surface of the obstacle known as "drag."

Now, if we place in the air stream an object so shaped (streamlined) that air pressure flows smoothly around it, the *pressure is* more nearly even on all sides. By altering the shape of this stream-lined object just a little, we can change relative pressures of air on its surface to produce "lift," and reduce "drag."

Lift is produced by *differences* in the pressure of air above and below a wing. The Swiss mathematician Daniel Bernoulli published a scientific paper in 1738 titled, *Hydrodynamica*, which discussed the forces of fluids in motion. One part explained how pressure of a moving fluid changes with motion. The same principle applies to gases, or air. In a teardrop-shaped wing, the speed of air passing over its top and bottom would be identical, but if we cut it in half, lengthwise, the basic wing shape of an airplane results. With this shape, air molecules moving over the top of the wing, or the curved upper surface, would have farther to go before reaching the back of the wing. They would need to move faster to keep up with the air molecules flowing along the flat bottom of the wing, where they have a shorter distance to travel to reach the wing's rear edge. The speed-up of the air flow over the top of the wing results in a drop in air pressure, which applies *suction* or *lift* to the top of the wing.

While most of the lift on a bird's wing or an airplane's wing comes from the low air pressure on its top, a certain amount of lift is generated from beneath by air molecules striking the undersurface of the wing. Here the effect is the reverse of what happens when air flows over the top. Air flow on the underside is stopped at a point close to the wing's front, or leading edge; then it gradually speeds up until it is near the back, or trailing, edge by which time it has reached the same velocity of the air traveling over its upper surface. In accordance with Bernoulli's principle, the slowed air along the bottom of the wing is at a higher pressure than the air speeding over the top of the wing and creates an upward pressure or lift below the wing.

Increasing the Lift–If the front edge of our newly designed wing is tilted upward just a little and it is placed in an air stream, the air will strike the bottom surface more directly and the lifting force on the wing from below is increased. The more the wing is tilted upward, the more lift it will get—up to a certain point. As the angle of tilt of the wing approaches the vertical, the air pressure against the bottom surface begins to push it *backward* rather than upward. Eventually, if the airplane wing is tilted too much, the lifting force vanishes and the drag is so great that it stops the plane's buoyancy or forward movement. This results in what we call a "stall." The plane must regain the proper wing angle and speed of its flight or it will crash.

Stalling Effect on Birds – The best resultant force (or lift above the wing) for the flight of a bird is when the "angle of attack" (angle of wing tilted up from horizontal) is between 3 degrees and 5 degrees. As the angle of attack of the bird's wing is increased, the center of pressure and lift moves forward from back to front of wing, and at the same time increases in strength, but begins to pull *less* upward and more backward until at an angle of attack of 25 degrees, the wing, or airfoil, is not flying at all and stalls because of turbulent eddies of air in back of it. This stalling because of the air reversal on top and under a bird's wing, lifts up the bird's wing coverts (the "stall feathers") on any bird just before it lands. Presumably the bird feels this and is warned that it is about to lose its wing lift altogether.

If a bird is planning to alight, it raises the angle of attack of its wings higher and higher. This makes the bird's forward movement become slower and slower until the wings stall and the bird ceases to fly. This point is the bird's "stalling speed" and for each kind of bird there is a speed through the air and an angle of glide beyond which it cannot fly. Many birds in alighting do this deliberately, and end with their wings raised or displayed (unfurled), at which the stall comes a few inches above the ground. Their legs and feet, held downward, absorb the shock as the bird settles gently to the ground.

A Bird's Propellers – Once the wing has been streamlined for flight (see under *Design of Wing*, second paragraph), the next step is to move it through the air fast enough to generate *lift*. (See previous section, *Lift.*) In an airplane this is accomplished by equipping it with propellers which are actually another set of wings whose lift is exerted forward rather than upward. This is basically, then, a single mechanism which, placed in one position holds the airplane up, and in another position drives it forward.

In the bird we find exactly the same mechanism of a "lifting wing" and a driving propeller built in each of the bird's wings and used in the same two ways as the plane's *separate* wings and propellers.

The wing of a bird consists of two functional parts—an inner part, nearest the body and operated from the shoulder joint; and an outer part which is moved separately by the bird's wrist which is about midway along the wing. The inner part of the wing gives lift almost exclusively. It is held rather rigidly in a slight tilt forward like the wing of an airplane. It also has the streamlined shape of a plane's wing: its upper feathers on top of the wings are curved back over the top of the wing in a curved surface. At the front edge of the bird's "wrist," where the inner and outer wings join, is a small group of feathers called collectively the *alula,* or "bastard wing." This is the bird's auxiliary airfoil to help it to gain additional lift when taking off from the ground or in landing. The bird can raise its alula to form an open slot between the alula itself and the main wing. Without the alula a bird cannot alight or take off from land successfully.

Each bird has a pair of "propellers" seen best in slow-motion photography of the bird in flight. During the downward beat of the bird's wings, the primary, or flight, feathers at the wing tips, stand out almost at right angles to the rest of the wing, and to the bird's line of flight. They take this form for only a split second during each wingbeat. *But this ability of a bird to change the slope and position of its wings is the key to bird flight.* Throughout the entire wingbeat, the flight feathers are constantly changing their shape, along with the wings, and adjusting "automatically" to air pressure and the changing flight requirement of the wing as it moves up and down.

This automatic adjustment is possible because of special feather design. The front vane of a wing-tip feather (a primary) on the forward side of the quill is much narrower than the rear, broader part of the vane. From this difference the force of the air twists the feathers into the shape of a propeller. As the wings beat downward against the air, the greater pressure against the wide rear vane of each of these primary feathers twists the vane *upward* until the feather takes on the proper shape and angle to function as a propeller.

The degree and shape of the twist is controlled largely by the design of the quill, which is rigid at its base, flattened and flexible toward its outer end. The specialized design (evolution) of flight feathers is beautifully adapted to meet the varied demands of bird flight. The airplane propeller rotates in one direction around a pivot; the bird's propeller oscillates rapidly up and down, and it must automatically adapt its shape, position, angle, and speed to the changing requirements of the moment. The flight feathers are not fastened immovably to the bone of the wing, but are held by a broad, flexible membrane, which allows considerable freedom of movement of each feather.

While a bird is flying easily, only the tips of the primary feathers twist to become propellers. But if a bird is in a hurry, and beats its wings strongly against the air, the whole outer section, or the "hand" wing, from the wrist out, may be twisted by the greater pressure on the air into one big propeller.

Path of the bird's propellers is *downward and forward* on the downstroke; *upward and backward* on the upstroke. The amount of forward and backward motion varies with the wingbeat. When a bird beats its wings rapidly, as in taking off, the increased pressure on the air drives the wings forward on a more nearly horizontal path. In leisurely flight, the wings are beaten more nearly vertically.

The inner part of the bird's wings, when maintained, or held at a proper angle, supports the bird's weight in flight through the entire wingbeat. The angle of this wing to the horizontal (called the angle of attack) is constantly adjusted by the bird to maintain a steady lifting force.

A Bird's Motor – In free flight, the bird's powerful breast muscles (the pectorals) sweep the whole wing up and down from the shoulder. The inner part of the wing does not actually need to move, but it *acts as a handle* to move the propeller (outer part of wing), and it gives the propeller greater speed and power.

How a Bird Flies – In its normal flapping flight, the wings of a bird sweep through the downbeat fully extended. At the end of the downstroke, the bird flexes its "wrists" to begin the upsweep of the wings. The whole wings then start upward, the hand section first, with the primary feathers partially separated like the fingers of a hand, and the inner half of the wings pulling along after the outer or hand section. At the end of the upward sweep of the wings, the hand section, in a sudden powerful burst, flaps *up* and *out*, and resumes its position for the next downward sweep of the wings. The upstroke requires less time than the downstroke from beginning to completion.

A bird has almost complete control over the outer section of its wings, or the "hand section," just as a man has over his hands and fingers. During the evolution of the bird wing from a lizardlike ancestor's "hands," or front feet, two of the "fingers" or finger-bones have fused into one, and the others have disappeared. However, the big primary feathers have replaced them so completely that the bird has gained many more "fingers" and muscles than it lost during the evolutionary adaptation to flight. The bird can twist its "hand" (the outer wing primaries) to any position, waggle them about, and can even clap the "hands" (outer wings) together behind its head and in front of its breast—examples: pigeon, and ruffed grouse. That is why wing motion is so variable and difficult to understand.

The powerful downstroke that lifts and propels the bird off the ground is also a forward stroke, so much so that the wings often touch each other in front of the breast (a pigeon "claps" its wings together in quick take-off) and in most birds almost always come close to slapping together at take-off and in climbing. The downstroke compresses together the feathers of the whole length of the bird's wings, *each feather grabbing its full hold on the air.*

At the beginning of the *upstroke* with the twist of the wrist to bring the wings back and upward, this lifts first the "wrist," then the half-folded wing (outer wing) and swivels the feathers apart like slats in a Venetian blind, which lets the air slip by. The wing's upstroke also plays a part in driving the bird upward and onward. The quick flip of the wings in recovering from the downstroke and sweeping upward (accomplished by the snap of the wrists) has much less power but is still a part of the *sculling* motion best illustrated by the rotor screw action of a helicopter's big propellers—forward and down, backward, up, and around—a figure-eight movement.

In the 1860s, Etienne J. Marey, a French scientist, who experimented with studies of the way that insects fly, also experimented with studies of the wing movements of crows in flight. He attached bits of white paper to the wing tips of a crow and released the bird so that it could fly against a black background. His photographic image of the white tips of the crow's wings in flight showed a series of loops and undulations on the downbeat that swept *forward and downward,* made a loop at the bottom of the downstroke, then on the upbeat traveled *upward* and *backward* until the wing tip curved again at the high point of the stroke to make the next downbeat. This experiment refuted the theory previously held that in flight, birds used their wings as one rows a boat, pushing backward and downward with a return of the arms and oars of forward and upward.

The heavier the bird, the faster it must fly to stay aloft. Sir D'Arcy Thompson, an eminent British scientist, made a theoretical calculation (using linear dimensions of the house sparrow which weighs about ⅔ of an ounce, and an ostrich of 250 pounds) that proved that if an ostrich could fly, it would need to do so at a speed of 100 miles an hour, as minimal speed, to stay aloft. The average minimal speed of birds in order to stay aloft is approximately 16½ feet per second, or about 11 miles an hour.

Gliding Flight – Gliding is the simplest and most elementary form of flight and is practiced by all birds except possibly hummingbirds. In all likelihood, gliding was the original form of bird flight and it is still the simplest.

A bird's wings make no propulsive movement in gliding. Actually the bird is coasting "downhill" in relation to the flow of air. It is the form of flight used by an airplane in coasting to a landing. The two main forces acting on a bird in flight are the pull of gravity and the resistance of the air to its passage. A gliding bird, coasting downward, is simply using its weight to overcome the air resistance to its forward motion.

The gliding bird's source of power is then in gravity, or the bird's weight. Without the earthward pull of gravity, the bird's wings could not move it through the air. By contrast, in flapping flight, the bird's wingbeating overcomes gravity and sets up its own air flow by movements of the wings. This creates the "lift," or resultant force that overcomes the pull of gravity and keeps the bird in flight.

Gliding in still air means a loss of height to a bird. In order to regain height the bird (if it does not wish to flap its wings, or start up its "motor") must find rising currents of air on which to glide (called static soaring), or by using adjacent air currents that are rising at different velocities (called dynamic soaring).

Soaring Flight – A soaring bird is one that maintains or even increases its altitude without flapping its wings. The three main requirements for a bird to soar successfully are large size, light wing loading (the weight of the bird in pounds borne by each square foot of its wing surface), and maneuverability in the air, especially the ability to make small, or tight, circles, or to change the direction of flight suddenly to take advantage of changes in the flow of the air.

Large size gives a bird sufficient momentum to carry it through small erratic air currents without loss of its stability or control of its flight. However, this stability of flight is a hindrance to maneuverability, and excellence in both is not usually found in one bird. The relatively small wing muscles of the soaring birds—vultures, red-tailed, red-shouldered, and broad-winged hawks for example—shows that this is a type of flight economical of a bird's efforts or energy. Vultures that soar so much over land, and albatrosses over the oceans, have become so highly adapted to soaring that they use this method of flight almost exclusively.

Soaring Flight of Land Birds – The golden eagle, the California condor, the turkey vulture, and the red-tailed hawk are some of the typical highly successful soaring birds that live on land and especially in open areas where they may be seen soaring high in the sky. These soaring land birds, the "static soarers" that keep aloft mainly by riding rising air currents, have evolved a type of wing that one scientist calls the Slotted-Soaring, or High-Lift Wing. The birds with this wing type are most often seen rising in spirals higher and higher in the sky on warm updrafts of air called "thermals." (See chapter IV.)

Maneuverability is especially important to the land soarers because the thermals on which they are carried upward are often small and undependable pillars of warm air. To circle in tight spirals within these small columns of air, the soaring birds need short but broad wings. The short wings give its flight low inertia and a quick, sensitive response to the unpredictable air currents; the broad wings with their slotting of the primary feathers, give it the high lift capacity, which is also helped by a high camber of the wings of the soaring bird (camber is the arching of the wings).

The soaring land birds are *slow fliers* because the shape of their wings is adapted to soaring. To soar in tight spirals a bird cannot fly at high speed. For it to fly slowly and yet to gain enough lift for its wings to avoid sinking, requires that the wings be tilted quite high in their "angle of attack." To do this without stalling is the problem of the soaring landbird.

The efficient, deeply slotted primary feathers of the wing tips of vultures, eagles, and the soaring hawks makes possible their combination of slow speed and high lifting power of their wings. As the vulture, hawk, or eagle soars upward, its primary, deeply slotted wing feathers are spread wide like the fingers of a hand. Each separately extended primary feather acts as a narrow wing with high lift capacity and set at a very high angle of attack (raised or slanted high from the horizontal).

To suit the demands of the air at any moment, a soaring bird alters the shape and expanse of its wings by either fanning (spreading) or folding its wing feathers; by changing its wingspread camber, or arching; the sweepback of its wings; or their angle of attack. By its ability to make these adjustments, a bird can select the best lift-to-drag ratio for any given speed of its flight.

Obstruction Air Currents – Besides the thermals, or so-called convection currents of air, soaring and gliding birds use *obstruction currents* which are updrafts of air caused when a steady or prevailing wind strikes and rises over such objects as mountains, hills, buildings, sand dunes, or even ships at sea, and ocean waves, which deflect the wind upward. A dramatic example of the use by birds of obstruction currents of air may be seen at Hawk Mountain Sanctuary near Kempton, Pennsylvania. Ornithologists and bird watchers in general who have traveled to Hawk Mountain each fall since the sanctuary was established in 1934, call the ridge the birds' "glider highway." (See chapters V and VI.)

Soaring and Gliding Flight of Seabirds – Although thermals, or warm currents of air, rise up over the oceans as well as over land, they are not as consistently available over the water, therefore they are too unreliable for their dependent use by the larger seabirds. Thermal updrafts rise from the water only when the air temperature is colder than the body of water below. The cold air, warmed by the water, rises. This is likely to occur in winter over the North Atlantic Ocean. When it does, gulls may be seen far from land, soaring high in the air, riding a whole group of the thermals, or columns of air, that are packed together like cells in a honeycomb. If the wind freshens, it may blow these columns of warm air over until they lie horizontally above the water. Then the flat-lying columns may rotate around their axes, each in the *opposite* direction from its neighbor. This rotation pushes up between them a ridge of rising air on which gulls and other birds can then glide over the surface of the ocean in a straight line.

If a warm front of air should move in over the ocean, bringing air that is warmer than the sea water, then the convection or rising warm air currents will stop. The gulls can then soar only by using obstruction, or "bounce" currents of air from wind striking the waves. If there is no wind, the gulls to stay aloft will be forced to flap their wings, which results in flapping flight instead of soaring flight.

Although shearwaters, petrels, and other small oceanic birds use the rising obstruction currents of air from wind striking waves or ocean swells, and gulls use the rising obstruction currents of air from wind striking the sides of ships, these deflected updrafts are thought to be too small and irregular and too near the surface of the water to sustain a large soaring bird such as an albatross.

Dynamic Soaring of the Albatross—Albatrosses, some of which have the widest wingspan of any seabird in the world, are capable of flapping flight, but it is as a glider, or living sailplane that this remarkable bird is able to course for weeks or months over the seas in its search for food on the ocean's surface. Because of its long, narrow glider-type wings, the albatross is not an efficient flying, or flapping, bird and must have a reliable source of energy without resorting to wing flapping. Because it cannot depend on the unreliable thermals, and the obstruction currents from waves that are used by the smaller shearwaters and petrels, the albatrosses must live almost exclusively in the southern oceans between the Tropic of Capricorn, south to the Antarctic Circle. There they find the source of energy they need. It is in the steady winds that blow continuously in those latitudes, from west to east. And it is on these winds that the albatross practices its highly specialized type of flight known as dynamic soaring.

The westerlies, or trade winds, blowing across the ocean's surface, become "stacked up" at various speeds. Owing to the friction at the ocean's surface, the lowest winds, or those traveling across the water itself, are slowest. As a result, wind speeds vary considerably from the ocean's surface upward, and it is because of this variation that albatrosses can practice sustained flight. At the ocean's surface the wind may be blowing only 20 miles an hour; 10 feet above the surface it may be blowing 32 miles an hour; at 20 feet, 36 miles an hour; at 50 feet, 40 miles an hour. (See chapter VII.)

Hovering Flight–Many kinds of birds can hover, that is, hold themselves in mid-air by flapping their wings sufficient to hold their position over one point on the ground or water below them. Some kinds of hawks and kingfishers are particularly expert at it as they hover in the air to watch for their prey directly below them. By beating its wings up and down and depressing and fanning its tail, with its body held almost vertically, a kingfisher or a sparrow hawk holds its position in the air against the downward pull of gravity. There it poises on beating wings until it is ready to dive downward— the kingfisher, with its wings held close to its sides to lessen the air resistance, in its plunge into the water to seize a minnow in its bill; the sparrow hawk, with folded wings, to dive to the ground and seize a grasshopper or a mouse with its talons.

Of all birds, the small hummingbirds are the most skillful at hovering. In their method of hovering before flowers to feed, their wings have become especially adapted to a helicopterlike flight. The hummingbird's flight method has been likened to the circular whirl of the helicopter's rotor, which allows it to hover, or to move ahead, backward, or sidewise at will. (The hummingbird is the only bird that can fly backward.) If the helicopter hovers, the rotor is in a plane parallel to the earth's surface—so are the wings of a hummingbird. As the helicopter moves forward or backward, the rotor tilts in the appropriate direction—so do the wings of a hummingbird. The helicopter can rise from a given spot without a runway for take-off—so can a hummingbird.

Hovering flight has made such demands on the wings of humming-birds that drastic changes in its wing structure, from that of other birds, have evolved. In its bony structure, the wing is practically all hand. The bones of the upper arm and forearm, which are relatively long in a bird like the pelican, are short and rigid in the hummingbird. However, the hummingbird's hand section of the wing is relatively long compared with that of the pelican. The elbow and wrist joints of the hummingbird are practically rigid. They make a permanently bent, inflexible framework that the hummingbird can move only from the shoulder, but very freely and in almost any direction.

The hummingbird is a speedster—its wings are rather flattened and taper to a relatively slender, elliptical tip, with a pronounced sweepback of the forward edge of the wings, much like that of the swallows, chimney swifts, and falcons. In proportion to its body size, the hummingbird's wings are average. They are thin, flat, and pointed with no slotting as in the soaring birds and many others. Apparently the wing is essentially a variable pitch propeller.

The Dive—A bird diving through the air is executing nothing more than a steep glide. It is usually done with the wings drawn in and back to lessen the air resistance, and the legs tucked up in the belly feathers. On motionless wings, it descends in its vertical plunge. A good trained peregrine falcon, 400 to 500 feet above a grouse that has been flushed below her, turns over in the air, gives two or three swift, powerful beats of her wings downward, then stoops, with her wings partly retracted, or drawn in toward her sides. The peregrine's stoop, dive, or plunge, is usually accomplished in about three to four seconds from a height of 400 to 500 feet above the ground. While she is stooping, the air is flowing over the falcon's retracted wings and she may move them a little as she plunges downward. Even though the wings are retracted, the rush of air under and over the wings tends to exert "lift" which flattens out the peregrine's dive, tending to pull her nose and body up toward the horizontal. The dive may start vertically, but after 200 to 300 feet of descent the air tends to pull the bird from a straight downward dive to one that is about 60 or 70 degrees from the vertical.

The Pitch-up, or "Throw-up"—Upward Glide—The pitch-up or "throw-up" is the falconer's term for a bird performing an upward glide. The motive power for the pitch-up is usually the carry over of momentum from the speed of its dive. A peregrine falcon often pitches upward at the bottom of its stoop and will bound high in the air to reach almost the same height from which it started. In this way, by pitching up, it can dive again and again on its prey. Also, a swallow after flying low over the ground will sometimes shoot straight upward in a vertical climb, its wings held motionless back along its sides.

Flight Maneuvering—The body of a bird is compact, with the heavier parts grouped closely around the center of gravity and set below the horizontal line of its wings. The effect of the weight of the bird during flight, as it is "suspended" between the wings, is somewhat like that of a pendulum. When gliding, vultures and other soaring birds often hold their wings at a "dihedral" angle, or slanted upward. This increases the bird's stability and keeps the pendulumlike body weight from swinging from side to side. What it accomplishes is to shift the centers of air pressure (called the resultant force) on the wings even higher above the center of gravity of the bird which prevents it from turning over on one side or the other. (See chapter IV.)

If a bird wants to increase its maneuverability, it sets its wings at an angle below the level of its body in a position called "anhedral." Swallows and swifts, birds of rapid flight that pursue flying insects in the air, often put the wings in this position as it enables them to twist and turn much more quickly in the air in the pursuit of their prey. The anhedral position of the wings, in which they are held lower than the lateral axis of the bird, is often used along with movements of the bird's tail to increase its maneuverability.

Mounting and Ringing—When a bird flies straight upward and forward, with its head higher than its tail, it is said to be "mounting." Birds are said to be "ringing" when they fly upward in a spiral, which they are likely to do when pursued by a hawk in order to gain height and thus escape.

Formation Flying—(See chapter IV.)

Mass Flights—Maneuvers of the Flock. (See chapter XI.)

Aerobatics—Looping—Some birds occasionally turn over in flight through a complete forward turn like the rolling of a wheel over the ground. The purpose of these loops may be to escape a pursuer or in pursuing, to gain an advantage over the prey. Some birds loop apparently in a spirit of play, or for the joy it seems to give them.

The Greenland gyrfalcon, the European hobby (a hawk similar to our American pigeon hawk), the marsh hawk, some of the large soaring hawks, ravens, and the peregrine, all have been seen to turn over in a loop several times in succession.

TABLE I

APPROXIMATE WINGSPANS, WEIGHTS, AND BODY LENGTHS OF
SOME OF THE LARGER BIRDS OF NORTH AMERICA, AND INCLUDING
THE WANDERING ALBATROSS OF THE SOUTHERN OCEANS

NOTE: Only the relative approximate sizes are shown, because wingspans, weights, and lengths will vary in individual birds. Also, the females of birds of prey (eagles, falcons, hawks, and owls) are up to about one third larger than the males, whereas the males of waterfowl (swans, geese, ducks) are usually larger than the females. Most figures given in the table are averages. Authorities cited in the right-hand column, and their publications from which the figures given here were taken, are listed in the Bibliography.

Bird	Wingspan[1]	Total Weight (lbs.)	Total Length[2]	Authority
Wandering albatross	11½ ft. max.	26¾ (male) 15–20	4½ ft.	William Jameson Robert Cushman Murphy
California condor	9 ft. av. 9 ft. 7 in. max.	20 (av.)	4 ft. 4½ in.	Carl B. Koford

[1] Wingspan is the width of the wings, from tip to tip when outstretched and measured across the bird's back. Same as measuring a man's total reach when his arms are stretched out straight from his sides.
[2] Total length of a bird is from the tip of its bill to the end of its tail.

Bird	Wingspan	Total Weight (lbs.)	Total Length	Authority
Trumpeter swan[3]	7–8 ft.	28	5 ft.	Winston E. Banko
Mute swan[4]	7–8 ft.	50.6 27 (av.)	4½ ft.	Fisher & Peterson Richard H. Pough
Whistling swan	7 ft.	16	4 ft. 4 in.	Richard H. Pough
White pelican	8½–10 ft. Up to 9 ft.	17 15–17	4½–5½ ft. —	Florence M. Bailey J. Stokley Ligon
White-tailed sea eagle	7–7½ ft.	15	3–3½ ft.	Richard H. Pough
Steller's sea eagle	6½–7½ ft.	12–14	3 ft.	Richard H. Pough
Frigate bird	7½ ft.	3½	3½ ft.	Richard H. Pough
Golden eagle	6½–7 ft.	9–12	33–39 in.	Richard H. Pough
Bald eagle	6½–7 ft.	9–13	34–36 in.	Richard H. Pough
Whooping crane[5]	7½ ft. (5 ft. tall)	8–10	4½ ft.	John J. Audubon & Richard H. Pough
Brown pelican	6–7 ft.	8¼	4 ft. 2 in.	Richard H. Pough

[3] The trumpeter swan is the largest native North American waterfowl.
[4] The mute swan is an introduced European bird.
[5] The whooping crane is the tallest North American bird.

Bird	Wingspan	Total Weight (lbs.)	Total Length	Authority
Sandhill crane	6½ ft.	8–13	3½–4 ft.	Richard H. Pough
Great white heron	6 ft. 3 in.	?	4 ft. 1 in.	Richard H. Pough
Great blue heron	5 ft. 10 in.	7	3 ft. 10 in.	Richard H. Pough
Gannet	6 ft.	?	3 ft.	Richard H. Pough
Turkey vulture	5 ft. 10 in.	3½	2½ ft.	Richard H. Pough
Black vulture	4 ft. 9 in.	4½	2 ft. 1 in.	Richard H. Pough
Canada goose[6]	3 ft. 7½ in. to 6 ft. 4 in.	2½ to 14	1 ft. 10 in. to 3 ft. 3½ in.	Richard H. Pough
Osprey	5 ft. 8 in.	3½	1 ft. 1 in.	Richard H. Pough
Great black-backed gull	5 ft. 5 in.	?	2 ft. 5 in.	Richard H. Pough
Blue-faced booby	5 ft. 3 in.	4¾	2 ft. 8 in.	Richard H. Pough
American flamingo	5 ft.	7	4 ft.	Robert P. Allen

[6] There are many races, or "forms" (subspecies) of the Canada goose, some no larger than a mallard duck. Others may have a wingspread of more than 6 feet and weigh 14 pounds or more. These forms nest in different parts of North America, principally in Canada.

Bird	Wingspan	Total Weight (lbs.)		Total Length	Authority
Wild turkey	5 ft.	9–10 { adult females 12–20 { adult males		3–4 ft.	Mosby and Handley
Glaucous gull	5 ft.	?		2 ft. 4 in.	Richard H. Pough
Great gray owl	5 ft.	?		2 ft. 6 in.	A. C. Bent
Great horned owl	4½ to 5 ft.	3		1 ft. 6 in. to 2 ft. 1 in.	Austing and Holt
Snowy owl	4½ to 5 ft.	?		2 ft. 1 in.	Pough and Bent
Skua, or great skua	4 ft. 11 in.	2½		1 ft. 9 in.	Richard H. Pough
White-fronted goose	4 ft. 10 in.	5½		2 ft. 5 in.	Richard H. Pough
Snow goose	4 ft. 10 in.	6		2 ft. 5 in.	Richard H. Pough
Herring gull	4 ft. 8 in.	2½		2 ft.	Richard H. Pough
Ring-billed gull	4 ft.	?		1½ ft.	Richard H. Pough

TABLE II

NUMBER OF WINGBEATS PER SECOND OF SOME
NORTH AND SOUTH[*] AMERICAN BIRDS

Kind of Bird	Wingbeats Per Second	Authority and Method of Measurement
Calliphlox amethystina [*] (A South American hummingbird) male female	80 60	Crawford H. Greenawalt[1] (high-speed photography)
Ruby-throated } male hummingbird } female	70 50	
Black-capped chickadee	27	
Mockingbird	14	
Giant hummingbird[*] *Patagona gigas*	8–10	
American goldfinch	4.9	Charles H. Blake[2] (using a fifth-second stop watch)
Starling	4.3	
Ring-necked pheasant	3.2	
Eastern bluebird	3.1	
Domestic pigeon	3.0 (rate quite varied)	

[1] *Hummingbirds,* published by Doubleday and Co., Inc. 1960.
[2] "Wing-flapping Rates of Birds," *Auk,* October 1947, pp. 619–20. Dr. Blake collected records of wing-flapping rates of wild birds in their normal, or unhurried

Kind of Bird	Wingbeats Per Second	Authority and Method of Measurement
Double-crested cormorant	2.6	
Mourning dove	2.45	
Laughing gull	2.45	
Killdeer	2.4	Charles H. Blake[2] (using a fifth-second stop watch)
Belted kingfisher	2.4	
American sparrow hawk	2.4	
Robin	2.3	
Herring gull	2.3	
Yellow-shafted flicker	2.2	
Common crow	2.0	

flight, while they were flying at constant speeds. He found the fifth-second stop watch sufficiently accurate, or satisfactory, up to as high as seven or eight flaps a second for a bird. Dr. Blake's corrections of plus or minus fractions of a wingbeat per second are not shown.

COMMON AND SCIENTIFIC NAMES OF BIRDS MENTIONED IN THE TEXT

Both the common and scientific names listed are those used by the American Ornithologists' Union in the *Check-List of North American Birds*, Fifth Edition, 1957, and of *A Field Guide to the Birds of Britain and Europe*, by Roger Tory Peterson, Guy Mountfort, and P. A. D. Hollom. For hummingbirds the common and scientific names are those used by Crawford H. Greenawalt in his book *Hummingbirds*.

Albatross, Black-browed, *Diomedea melanophris*
    Black-footed, *Diomedea nigripes*
    Laysan, *Diomedea immutabilis*
    Short-tailed, *Diomedea albatrus*
    Wandering, *Diomedea exulans*
    Yellow-nosed, *Diomedea chlororhynchos*
Archaeopteryx, *Archaeopteryx lithographica*
Auk, Razorbill. See Razorbill
Auklet, Cassin's, *Ptychoramphus aleutica*
    Least, *Aethia pusilla*

Blackbird, Red-winged, *Agelaius phoeniceus*
Bobolink, *Dolichonyx oryzivorus*
Bobwhite, *Colinus virginianus*
Bunting, Indigo, *Passerina cyanea*

Cardinal, *Richmondena cardinalis*
Catbird, *Dumetella carolinensis*
Chickadee, Black-capped, *Parus atricapillus*
Chicken, Prairie, *Tympanuchus cupido* and *T. pallidicinctus*
Chough, *Coracia pyrrhocorax* and *C. graculus*
Condor, California, *Gymnogyps californianus*
Cormorant, Double-crested, *Phalacrocorax auritus*
Crane, Sandhill, *Grus canadensis*
Crane, Whooping, *Grus americana*
Creeper, Brown, *Certhia familiaris*
Creeper, Wall, *Tichodroma muraria*
Crow, Common, or American, *Corvus brachyrhynchos*

Dipper, or Water Ouzel, *Cinclus mexicanus*
Dove, Rock, *Columba liva* (domestic pigeon)
    Mourning, *Zenaidura macroura*
Duck, Oldsquaw. See Oldsquaw
    Pintail. See Pintail
    Wood, *Aix sponsa*
Dunlin, *Erolia alpina* (red-backed sandpiper)

Eagle, Bald, *Haliaeetus leucocephalus*
    Golden, *Aquila chrysaetos*
Egret, Reddish, *Dichromanassa rufescens*

Falcon, Peregrine. See Peregrine
Finch, House, *Carpodacus mexicanus*
    Purple, *Carpodacus purpureus*
Flicker, Red-shafted, *Colaptes cafer*
    Yellow-shafted, *Colaptes auratus*
Frigate Bird, *Fregata magnificens* (man-o'-war bird)

Goose, Blue, *Chen caerulescens*
    Canada, *Branta canadensis*
    Snow, *Chen hyperborea*
Goshawk, *Accipiter gentilis atricapillus*
Grebe, Pied-billed, *Podilymbus podiceps*
Grosbeak, Evening, *Hesperiphona vespertina*

Grouse, Ruffed, *Bonasa umbellus*
Guillemot, Black, *Cepphus grylle*
    Pigeon, *Cepphus columba*
Gull, Herring, *Larus argentatus*
Gyrfalcon, *Falco rusticolus*

Hawk, Cooper's, *Accipiter cooperii*
    Fish. See Osprey
    Marsh, *Circus cyaneus*
    Pigeon, *Falco columbarius*
    Red-shouldered, *Buteo lineatus*
    Red-tailed, *Buteo jamaicensis*
    Sharp-shinned, *Accipiter striatus*
    Sparrow, *Falco sparverius* (American sparrow hawk or kestrel)
    European Sparrow, *Accipiter nisus*
Heron, Green, *Butorides virescens*
Hummingbird, Allen's, *Selasphorus sasin*
    Anna's, *Calypte anna*
    Bee, *Calypte helenae*
    Black-chinned, *Archilochus alexandri*
    Broad-tailed, *Selasphorus platycercus*
    Calliope, *Stellula calliope*
    Costa's, *Calypte costae*
    Ruby-throated, *Archilochus colubris*
    Rufous, *Selasphorus rufus*
    Sword-billed, *Ensifera ensifera*

Jackdaw, *Corvus monedula*
Jay, Blue, *Cyanocitta cristata*

Kestrel. See American Sparrow Hawk
Killdeer, *Charadrius vociferus*
Kingbird, Eastern, *Tyrannus tyrannus*
Kingfisher, Belted, *Megaceryle alcyon*
Kinglet, Golden-crowned, *Regulus satrapa*

Lammergeyer. See Vulture, Bearded
Lapwing, *Vanellus vanellus*

Lark, Horned, *Eremophila alpestris*
Limpkin, *Aramus guarauna*
Lizard-Bird. See Archaeopteryx
Loon, Common, *Gavia immer*

Mallard, *Anas platyrhynchos*
Man-o'-War Bird. See Frigate Bird
Meadowlark, Eastern, *Sturnella magna*
     Western, *Sturnella neglecta*
Mockingbird, *Mimus polyglottos*
Murre, Common, *Uria aalge*

Nighthawk, Common, *Chordeiles minor*

Oldsquaw, *Clangula hyemalis*
Osprey, *Pandion haliaetus*
Owl, Great Horned, *Bubo virginianus*
     Short-eared, *Asio flammeus*
Oystercatcher, European, *Haematopus ostralegus*

Pelican, White, *Pelecanus erythrorhynchos*
Penguin, Humboldt's, *Spheniscus humboldti* (Peruvian penguin)
     King, *Aptenodytes patagonicus*
     Rockhopper, *Eudyptes crestatus*
Peregrine, *Falco peregrinus anatum*
Pheasant, Ring-necked, *Phasianus colchicus*
Pigeon, Domestic. See Rock Dove
Pintail, *Anas acuta*
Plover, Black-bellied, *Squatarola squatarola*
Puffin, Horned, *Fratercula corniculata*
     Tufted, *Lunda cirrhata*

Quail, Bobwhite. See Bobwhite
     California, *Lophortyx californicus*

Rail, King, *Rallus elegans*
Raven, Common, *Corvus corax*
Razorbill, *Alca torda*
Redbird. See Cardinal

Robin, *Turdus migratorius*
Rook, *Corvus frugilegus*

Sandpiper, Red-backed. See Dunlin
    Semipalmated, *Ereunetes pusillus*
    Spotted, *Actitis macularia*
Sapsucker, Yellow-bellied, *Sphyrapicus varius*
Snipe, Common, *Capella gallinago*
Sparrow, House, *Passer domesticus* (English sparrow)
    Savannah, *Passerculus sandwichensis*
    Golden-crowned, *Zonotrichia atricapilla*
Starling, *Sturnus vulgaris*
Stork, Wood, *Mycteria americana* (wood ibis)
Swallow, Barn, *Hirundo rustica*
    Cliff, *Petrochelidon pyrrhonota*
    Rough-winged, *Stelgidopteryx ruficollis*
    Tree, *Iridoprocne bicolor*
Swan, Mute, *Cygnus olor*
Swift, Alpine, *Apus melba*
    Chimney, *Chaetura pelagica*
    (European), *Apus apus*
    Spine-tailed, *Hirundapus caudacutus*

Teal, Cinnamon, *Anas cyanoptera*
Tern, Common, *Sterna hirundo*
Thrush, Varied, *Ixoreus naevius*
Turkey, Wild, *Meleagris gallopavo*

Vulture, Bearded, *Gypaetus barbatus* (lammergeyer)
    Black, *Coragyps atratus*
    Turkey, *Cathartes aura* (turkey buzzard)

Waxwing, Cedar, *Bombycilla cedrorum*
Whippoorwill, *Caprimulgus vociferus*
Woodcock, American, *Philohela minor*
Wren, Winter, *Troglodytes troglodytes*

BIBLIOGRAPHY

ALLEE, W. C.

1938. Co-operation Among Animals. Sir Isaac Pitman and Sons, Ltd., London.

ALLEN, ARTHUR A.

1961. The Book of Bird Life. D. Van Nostrand Co., New York.

ALLEN, F. H.

1939. Effect of Wind on Flight Speeds. *Auk*, 56:291–303.

ALLEN, ROBERT P.

1956. The Flamingos: Their Life History and Survival. Research Report No. 5, Nat. Aud. Soc., New York.

AMERICAN ORNITHOLOGISTS' UNION

1957. Check-List of North American Birds. Published by The Amer. Orn. Union, Lord Balt. Press, Inc., Baltimore.

AUDUBON, JOHN JAMES

1831–1839. Ornithological Biography. Published by author. Edinburgh.

AUSTIN, OLIVER L., JR.

1961. Birds of the World. The Golden Press, New York.

AUSTING, G. RONALD

1964. The World of the Red-tailed Hawk. J. B. Lippincott Co., Philadelphia.

AUSTING, G. RONALD, and HOLT, JOHN B., JR.

1966. The World of the Great Horned Owl. J. B. Lippincott Co., Philadelphia.

AYMAR, GORDON C.

1935. Bird Flight. Garden City Pub. Co., New York.

BANKO, WINSTON E.

1960. The Trumpeter Swan: Its History, Habits and Population in the United States. *North Amer. Fauna 63*, U. S. Gov. Printing Office, Washington.

BAILEY, FLORENCE MERRIAM

1928. Birds of New Mexico. New Mex. Dept. of Game and Fish, Sante Fe.

BARLEE, JOHN

1964. Flight. *In* A New Dictionary of Birds, Edited by A Landsborough Thomson. McGraw-Hill, New York.

BEEBE, C. WILLIAM

1906. The Bird: Its Form and Function. Henry Holt, New York.

BELLROSE, F. C., and GRABER, R. R.

1963. A Radar Study of the Flight Directions of Nocturnal Migrants. *Proc. of XIIIth Int. Orn. Cong.*, Vol. I:362–89, Ithaca, N.Y.

BENT, ARTHUR CLEVELAND

1919. Life Histories of North American Diving Birds. *U. S. Nat. Mus. Bul.* 107, U. S. Gov. Printing Office, Washington.

1922. Life Histories of North American Petrels and Pelicans and Their Allies. *U. S. Nat. Mus. Bul.* 121.

1927. Life Histories of North American Shorebirds. Part I, *U. S. Nat. Mus. Bul.* 142.

1932. Life Histories of North American Gallinaceous Birds. *U. S. Nat. Mus. Bul.* 162.

1937–39. Life Histories of North American Birds of Prey. Part I, *U. S. Nat. Mus. Bul.* 167, and *Bul.* 170, Part II.

1940. Life Histories of North American Cuckoos, Goatsuckers, Hummingbirds and Their Allies. *U. S. Nat. Mus. Bul.* 176.

1948. Life Histories of North American Nuthatches, Wrens, Thrashers and Their Allies. *U. S. Nat. Mus. Bul.* 195.

BLAKE, CHARLES H.

1947. Wing-flapping Rates of Birds. *Auk*, 56:619–20.

BRODKORB, PIERCE

1960. How Many Species of Birds Have Existed? *Bul. of The Flor. State Mus.*, Gainesville.

1963. Catalogue of Fossil Birds. Part I, *Bul. of the Flor. State Mus.*

BROOKS, ALLAN

 1945. The Underwater Actions of Diving Ducks. *Auk,* 62: 517–23.

BROUN, MAURICE

 1948. Hawks Aloft: The Story of Hawk Mountain. Dodd, Mead and Co., New York.

BROUN, MAURICE, and GOODWIN, BEN V.

 1943. Flight Speeds of Hawks and Crows. *Auk,* 60:486–92.

BROWN, R. H. J.

 1964. Flight. *In* Biology and Comparative Physiology of Birds. Edited by A. J. Marshall. Acad. Press, New York and London.

CARSON, RACHEL

 1951. The Sea Around Us. Oxford Univ. Press, New York.

COTTAM, CLARENCE, WILLIAMS, CECIL S., and SOOTER, CLARENCE A.

 1942. Flight and Running Speeds of Birds. *Wilson Bul.* 54: 121–31.

CRAIGHEAD, FRANK and JOHN

 1939. Hawks in the Hand. Houghton-Mifflin Company, Boston.

DEWAR, JOHN

 1924. The Bird as a Diver. H. F. & G. Witherby, London.

DRURY, W. H., JR., and NISBET, I. C. T.

 1964. Radar Studies of Orientation of Songbird Migrants in Southeastern New England. *Bird-Banding,* 35:69–119.

EARDLEY, A. J.

 1965. General College Geology. Harper and Row, New York.

FISHER, JAMES, and LOCKLEY, R. M.

 1954. Sea-Birds. Houghton-Mifflin Company, Boston.

FISHER, JAMES, and PETERSON, ROGER T.

 1964. The World of Birds. Doubleday and Company, Inc., Garden City.

FORBUSH, EDWARD I.

 1925–29. Birds of Massachusetts and Other New England States. Massachusetts Board of Agriculture.

GILLIARD, E. THOMAS

 1958. Living Birds of the World. Doubleday and Company, Inc., Garden City.

GLOVER, FRED A.

 1947. Flight Speed of Wild Turkeys. *Auk,* 64:623–24.

GRAHAM, R. R.
    1930. Safety Devices in Wings of Birds. *British Birds,* 24: 2–21; 34–47; 58–65.

REENAWALT, CRAWFORD H.
    1960. Hummingbirds. Doubleday and Company, Inc., Garden City.

IARTMAN, FRANK A.
    1961. Locomotor Mechanisms of Birds. *Smithsonian Miscellaneous Collections,* 143:No. 1. Washington.

HEILMANN, GERHARD
    1927. The Origin of Birds. D. Appleton and Co., New York.

HEINROTH, OSKAR and KATHARINA
    1958. The Birds. Univ. of Mich. Press, Ann Arbor.

HOCHBAUM, ALBERT
    1955. Travels and Traditions of Waterfowl. Univ. of Minn. Press, Minneapolis.

HUFFAKER, E. C.
    1897. On Soaring Flight. *Ann. Rep. of the Smith. Inst.,* 183–206.

IDRAC, P.
    1925. Le vol sans battlement des albatros ou comment volent les albatros. *La Nature,* premier semestre, 53 annee:241–47.
    1926. Le vol des albatros. *Rev. Franc. d' Ornith.* (2) 10: 38–46.

INGRAM, COLLINGWOOD
    1919. Notes on the Height at Which Birds Migrate. *Ibis,* 61: 321–25.

JACK, ANTHONY
    1953. Feathered Wings: A Study of the Flight of Birds. Methuen and Co. Ltd., London.

JAMESON, WILLIAM
    1958. The Wandering Albatross. Rupert Hart-Davis, London.

KELSO, J. E. H.
    1922. Birds Using Their Wings as a Means of Propulsion Under Water. *Auk,* 39:426–28.

KOFORD, CARL B.
    1953. The California Condor. *Research Report No. 4.* Nat. Aud. Soc., New York.

LACK, DAVID

1959. Migration Across the North Sea Studied by Radar. (Part I), *Ibis*, 101:209–34.

1960. Migration Across the North Sea Studied by Radar. (Part II), *Ibis*, 102:26–27.

1960. The Height of Bird Migration. *British Birds*, 53:5–10.

LANE, FRANK W.

1946. Birds vs. Planes. *Nat. Hist.*, 55:165.

1954. Nature Parade. Sheridan House, New York, New York.

LAWSON, RALPH

1930. The Stoop of a Hawk. *Bul. of Essex (Mass.) Ornith. Club*, 78–80.

LEWELLEN, JOHN

1953. Birds and Planes: How They Fly. Thomas Y. Crowell Company, New York.

LIGON, J. STOKLEY

1961. New Mexican Birds: And Where to Find Them. Univ. of N. Mex. Press, Sante Fe.

LOKEMOEN, JOHN T.

1967. Flight Speed of the Wood Duck. *Wilson Bul.*, 79:238–39.

MCCABE, T. T.

1942. Types of Shorebird Flight. *Auk*, 59:110–11.

MCILHENNY, E. A. and OSBORN, ROSEMARY

1938. Black Vulture Following Aeroplane. *Auk*, 55:521.

MCLEAN, DONALD DUDLEY

1930. The Speed of Flight of Certain Birds. *Gull*, 12, No. 3.

MCMILLAN, NEIL T.

1938. Birds, and the Wind. *Bird-Lore*, Nov.–Dec.

MEINERTZHAGEN, RICHARD

1920. Some Preliminary Remarks on the Altitude of Migratory Flight. *Ibis*, 62:920–36.

1955. The Speed and Altitude of Bird Flight. *Ibis*, 97:81–117.

MEEUSE, B. J. D.

1961. The Story of Pollination. The Ronald Press Co., New York.

MOODY, PAUL AMOS

1953. Introduction to Evolution. Harper and Brothers, New York.

MOSBY, HENRY S. and HANDLEY, CHARLES O.

 1943. The Wild Turkey in Virginia: Its Status, Life History, and Management. Com. Game and Inland Fisheries, Richmond.

MURCHIE, GUY

 1954. Song of the Sky. Houghton-Mifflin Company, Boston.

MURPHY, ROBERT CUSHMAN

 1936. The Oceanic Birds of South America. Macmillan Company, New York.

NEWMAN, B. G.

 1958. Soaring and Gliding Flight of the Black Vulture. *Journ. of Exper. Biolo.*, 35:280–85.

NISBET, I. C. T.

 1963. Measurements with Radar of The Height of Nocturnal Migration Over Cape Cod, Massachusetts. *Bird-Banding*, 34: 57–67.

NOPCSA, BARON FRANCIS

 1907. Ideas on the Origin of Flight. *Pro. of the Zool. Soc. of London.*

ODUM, EUGENE P., CONNELL, CLYDE E., and STODDARD, HERBERT L.

 1961. Flight Energy and Estimated Flight Ranges of Some Migratory Birds. *Auk*, 78:515–27.

PEARSON, OLIVER P.

 1953. The Metabolism of Hummingbirds. *Scient. Amer.*, 88: 69–72.

 1961. Flight Speeds of Some Small Birds. *Condor*, 63:506–7.

PENROSE, HARALD

 1949. I Flew with the Birds. Country Life Ltd., London.

PETERS, JAMES L.

 1931. Check-List of the Birds of the World, Vol. I. Cambridge University Press, Cambridge, Mass.

PETERSON, ROGER TORY

 1964. The Birds. Time and Life Inc., New York.

PITTMAN, JAMES A.

 1953. Direct Observation of the Flight Speed of the Common Loon. *Wilson Bul.* 65:213.

POOLE, EARL L.

 1938. Weights and Wing Areas in North American Birds. *Auk*, 55:511–17.

POUGH, RICHARD H.

    1935. A Glider Highway. *Bird-Lore,* Sept.–Oct.

    1946. Audubon Land Bird Guide. Doubleday and Company, Inc., Garden City.

    1951. Audubon Water Bird Guide. Doubleday and Company, Inc., Garden City.

RAND, AUSTIN L.

    1955. Stray Feathers from a Bird Man's Desk. Doubleday and Company, Inc., Garden City.

RASPET, AUGUST

    1960. Biophysics of Bird Flight. *Science,* 132:191–200.

RATHBUN, S. F.

    1934. Notes on the Speed of Flight of Birds. *Murrelet,* 15: 23–24.

RIVIERE, B. B.

    1922. Speed of the Domestic Pigeon. *British Birds,* 15:298.

ROMER, ALFRED S.

    1945. Vertebrate Paleontology. Univ. of Chic. Press, Chicago.

ROWAN, WILLIAM

    1931. The Riddle of Migration. Williams and Wilkins Co., Baltimore.

SAVILE, D. B. O.

    1950. The Flight Mechanism of Swifts and Hummingbirds. *Auk,* 67:499–504.

    1957. Adaptive Evolution of the Avian Wing. *Evolution,* 11: 212–24.

    1962. Gliding and Flight in the Vertebrates. *Amer. Zoologist,* 2:161–66.

SCHNELL, GARY D.

    1965. Recording the Flight-Speed of Birds by Doppler Radar. *The Living Bird,* Fourth Annual of Cornell Lab. of Orn., Ithaca.

SPIERS, J. MURRAY

    1945. Flight Speeds of the Old-Squaw. *Auk,* 62:135–36.

STAGER, KENNETH E.

    1964. The Role of Olfaction in Food Location by the Turkey Vulture (*Cathartes aura*). *Contributions to Science,* Los Angeles County Mus., Vol. 81.

STEVER, H. GUYFORD, and HAGGERTY, JAMES, and THE EDITORS OF
Life.
1960. Flight. Time and Life Inc., New York.

STILLSON, BLANCHE
1954. Wings: Insects, Birds, Men. Bobbs-Merrill Co., Inc., New
York.

STORER, JOHN H.
1952. Bird Aerodynamics. *Scient. Amer.,* 87:25–29.

STORER, ROBERT W.
1960. Evolution in the Diving Birds. *Proc. of XIIth Int. Ornith.
Cong. of 1958.* Helsinki, 694–707.
1960. Adaptive Radiation in Birds. *In* Volume I, Biology and
Comparative Physiology of Birds. Edited by A. J. Marshall,
Acad. Press, New York.

STUART-BAKER, E. C.
1922. Speed of Indian Swifts. *British Birds,* 15:31.
1942. The Speed of Birds. *Country Life,* March.

SWINTON, W. E.
1960. The Origin of Birds. *In,* Volume I, Biology and Comparative Physiology of Birds. Edited by A. J. Marshall. Acad.
Press, New York.

TERRES, JOHN K.
1946. Birds Have Accidents Too! *Aud. Mag.,* Jan.–Feb.
1946. Feathered Death in the Sky. *Coronet,* Aug.
1948. Smithsonian Bird Man: A Biographical Sketch of Alexander Wetmore. *Aud. Mag.,* May–June.

THORPE, W. H.
1956. Learning and Instinct in Animals. Methuen and Co.,
Ltd., London.

TOWNSEND, CHARLES W.
1909. The Use of Wings and Feet in Diving Birds. *Auk,* 31:234–
48.
1924. Diving of Grebes and Loons. *Auk,* 41:29–41.

VAN TYNE, JOSSELYN, and BERGER, ANDREW J.
1959. Fundamentals of Ornithology. John Wiley and Sons, Inc.,
New York.

WAKEFIELD, JOHN

1964. The Strange World of Birds. Macrae Smith Company, Philadelphia.

WALKINSHAW, LAWRENCE H.

1949. The Sandhill Cranes. Cranbrook Institute of Science, Bloomfield Hills, Michigan.

WALLS, GORDON LYNN

1948. The Vertebrate Eye and Its Function. Cranbrook Institute of Science, Bloomfield Hills.

WELTY, CARL J.

1955. Birds as Flying Machines. *Scient. Amer.*, 90:88–96.
1962. The Life of Birds. W. B. Saunders Co., Philadelphia.

WESTOVER, MYRON F.

1932. The Flight of Swifts. *Bird-Lore,* July–August.

WETMORE, ALEXANDER

1926. The Migrations of Birds. Harvard Univ. Press, Cambridge.
1950. Recent Additions to Our Knowledge of Prehistoric Birds. *Proc. of the Xth Inter. Orn. Cong.,* Uppsala.

WING, LEONARD W.

1956. Natural History of Birds: A Guide to Ornithology. The Ronald Press, New York.

WOODFORD, M. H.

1960. A Manual of Falconry. Charles T. Branford Company, Newton, Massachusetts.

ZUMBERGE, JAMES H.

1958. Elements of Geology. John Wiley and Sons, New York.

INDEX

Pterosaurs, 126
Puffins
    Atlantic, 120
    horned, 113, 156
    tufted, 113, 114, 115, 156
Purple finch, 80, 154
    and natural forces, 107

Quail, 72, 77
    bobwhite, 35, 110, 156
    California, 88, 156
    white muscles of, 72
Quill, 136

Radar, 80, 81, 92
    Doppler, 92
    and flight study, 77-78
    height finder, 77
    scope, 77
Rail, king, 156
Raspet, August, scientist, 28
    biophysics of flight, 28, 100
Ratio, span/width of wing, 55, 56
Raven(s)
    black, 31
    common, 156
    looping, 146
Razorbill, 119, 120, 153, 156
Records of flight speed, 80, 88, 89, 90, 92
Red-backed sandpiper(s), 87, 88, 154, 157
Redbird. see Cardinal
Reddish egret, 85, 154
Red flight muscles, 72
Red-shafted flicker, 89, 154
Red-shouldered hawk, 36, 48, 155
    soaring flight, 140
Red-tailed hawk, 23, *32*, 36, 38, *42*, 43, 44,
    45, 155
    courtship flight of, 91
    nesting, 43
    nesting territory, 44, 45
    soaring flight, 140
    sound of (of wings), 43
    speed of, 90, 91
    wingspan, 23, 41
Red-winged blackbirds, 153
    speed of, 92
Refuge, Canada, 2
Ring-billed gull, 150
Ringing, 145
Ring-necked pheasant, 156
    wingbeats of, 151
Robin, 152, 157
    and natural forces, 107
Rockhopper penguin, 116, 156
Rooks, 44, 157
    experiments of, 82
    high altitude record, 80
Rough-winged swallow, 97, 109, 157
Ruby-throated hummingbirds, 63-74, 155
    feeding of, 66
    flight, 65, 67
        motion pictures of, 67, 68
        speed of, 64, 68
    gorget of, 65, 66
    metabolism of, 72, 73, 74
    motion pictures, high speed, 67, 68
    nesting, 65
    and picture windows, 69-70
    wing, structure of, 70
    wingbeat, ratio of, 63, 68, 69, 151

Ruffed grouse, 105, 138, 155
    courtship sound, 101
    studies of diseased, 109
Rufous hummingbird, 155

Sandhill crane, 149, 154
    speed of, 92
Sandpipers, 70, 87, 88, 129
    leading bird of flock, 103
    red-backed, 87, 88, 154, 157
    semipalmated, 78, 102, 103, 157
    spotted, 157
        speed of, 92
Sapsucker
    and natural forces, 107
    yellow-bellied, 157
Savannah sparrow, 89, 157
Savile, D. B. O., scientist, 23, 36
Schlief, Emil, 106
Schnell, Gary D., 92
Schweizer Soaring School, 45
Sea parrots. See Tufted puffins
Sea pigeon. See Black guillemot
Semipalmated sandpiper(s), 78, 102, 103, 157
Shape of wings, 140, 141
    of albatross, 142
    and soaring flight, 141
    and wingbeat, 86
    See also Wings; Wingbeat; Wingspan
Sharp-shinned hawk, *32*, 34, 35, 38, 155
    speed of, 90
Shearwaters, 33, 107, 120, 142
    and obstruction currents, 142
Short-eared owl, 110, 156
    courtship flight of, 101
Short-tailed albatross, 59, 153
Shrikes, 88
Skua, 150
Slotted-soaring, 23, 36, 140
Smithsonian Institution, 89, 122
Snipe, 77
    common, 157
    sound of, 101
Snow geese, 154
    and lightning, 107
    wingspan of, 150
Snowy owl, 150
Snyder, Dorothy, ornithologist, 109
Soaring flight, description of, 140
    dynamic, *56*, 57, 58, 140, 143
        of albatross, 142-43
    gliding, 142
    and shape of wing, 141
    slotted, 23, 36, 140
    static, 33, 140, 141
Solnhofen limestone, 123
South American condor, 79, 80
    wingspread of, 80
Sparrow(s), 122
    golden-crowned, 81, 157
    house (English), 75, 110, 157
        speed of, 92
    savannah, 89, 157
    speed and weight of, 139
Sparrow hawks, 155
    American, 8, 26, 47, 48, 143
        wingbeats of, 152
    European, 82, 111, 155
        speed of, 90
Speed, 90, 91
    chickadee, 69

175

common crow, 69
Humboldt's penguin, 117, *117*
hummingbirds, 64, 68
Indian crow, 69
lift-to-drag ratio, 141
mockingbird, 69
penguins, 116, 117, *117*
pigeon, 10, 69
radar, 80, 81, 92
  Doppler, 92
  and flight study, 77–78
  height finder, 77
  scope, 77
  records of, 80, 88, 89, 90, 92
  and reserve, 86
  sinking, 28
  underwater, 116
  wandering albatross, 56
  and weight, 139, 145
  See also Wing; Wingbeat; Flight
Spine-tailed swifts, 157
  speed of, 92–93
Spotted sandpiper, 157
  speed of, 92
Sprunt, Alexander, Jr., ornithologist, 106
Stall, 135, 141
  feathers, 135
Starling(s), 79, 89, 108, 157
  wingbeats of, 151
Static soaring, 33, 140, 141
Steller's sea eagle, wingspan of, 148
Stilt-legged hawks, petrified, 123
Stoddard, Herbert L., 35
Stork(s), 79
  Maribou, 51
  wood, 17, 157
Storm petrels, 59
Structure of wing. *See* Wing
Stuart-Baker, E. C., British ornithologist, 92, 93
Swallow(s) 70, 75, 81, 82, 87, 88, 105, 129, 131, 144, 145
  barn, 38, 85, 157
  cliff, 85, 157
  rough-winged, 97, 109, 157
  speed of, 92
  tree, 85, 109, 157
Swan(s) wild, 46, 78, 79, 114, 122, 147
  flight height, comparison, *76*
  mute, 157
    music of wings of, 102
    wingspan of, 148
  trumpeter, 148
  whistling, 148
  wings of, 105
Swift(s), 81, 87, 145
  alpine, 85, 157
  chimney, 70, 82, 157
    speed of, 92
  European, 81, 82, 157
  high-altitude record, 80
  spine-tailed, 92–93, 157
Sword-billed hummingbird, 67, 155

Teal, cinnamon, 88, 157
Terns, common, 157
  speed of, 92
Tertiary ages, 126
Texas, hunting of golden eagles, 46–47
Thecodont reptiles, 125

Thermals, 23, 28, 33, 41, 43, 44, 45, 140, 141, 142
  and vultures, 28
Thompson, Sir D'Arcy, British scientist, 139
Thrashers, 75
Thrush, 75
  and natural accident, 109
  varied, 157
Tit, European, 82
Trade winds, 143
Tree swallow, 85, 109, 157
Trumpeter swan, wingspan of, 148
Tubenoses, 59
Tufted puffins, 113, 114, 115, 156
Turkey(s), wild, 60, 75, 150, 157
  Ice Age, petrified, 123
  speed of, 91
Turkey buzzards. *See* Turkey vultures
Turkey vultures, 21, *22*, 22–28, 36, 41, 43, 44, 79, 140, 141, 145, 149, 157
  soaring flight, 140
  speed, 90
  V-shaped flight, 21, *22*, *22*
  wingspan, 23, 149

U. S. Fish and Wildlife Service, 80, 130
  Bird-banding Office, 118
  Denver Wildlife Research Center, 81
Upper Jurassic period, 124

V-shaped flight, 21, 22, *22*, 106
Varied thrush, 109, 157
Vulture, 28, 31, 34, 36, 38, 43, 44, 48, *55*, 77
  bearded (*see* Lammergeyer)
  black, 27, 28, 149, 157
  flightless running, 123
  giant, petrified, 123
  southern black, 27, 43
  turkey, 21, *22*, 22–28, 36, 41, 43, 44, 79, 140, 141, 145, 149, 157
    soaring flight, 140
    speed of, 90
    V-shaped flight, 21, *22*, *22*
    wings of, 105
    wingspan, 23, 149

Walkinshaw, Lawrence H., ornithologist, 92
Wall creeper, 79, 154
Wandering albatross, 51–60, *56*, 153
  attacks by, 53–54
  common names, 53
  dynamic soaring, *56*, *57*, *58*
  and Eocene Age, 55
  nesting, 55
  as pets, 60
  speed of, 56
  wingspan, 51, 54, 55, 147
    ratio of, 55, 56
Warblers, 75, 122
Waterfowl, 2, 129, 131
Water ouzel. *See* Dipper
Waxwings, cedar, 92, 157
Weight, total, 147–50
Weight and flight speed, 139–40, 145, 147–50
Weitnauer, Emil, 81
Weller, David, 45
Western meadowlarks, 156
Wetmore, Alexander, ornithologist, on fossil birds, 122, 123, 126
  list of flight speeds, 89
Whippoorwill, 102, 157